DEC 2 9 2014
WITHDRAWN
D0435189

GOD'S PLANET

GOD'S PLANET

Owen Gingerich

Harvard University Press

Cambridge, Massachusetts
London, England
2014

Printed in the United States of America
First printing

Library of Congress Cataloging-in-Publication Data

Gingerich, Owen.
God's planet / Owen Gingerich.
pages cm
Includes bibliographical references and index.
ISBN 978-0-674-41710-6 (alk. paper)
1. Religion and science. 2. Copernicus, Nicolaus, 1473–1543. 3. Darwin, Charles, 1809–1882. 4. Hoyle, Fred, 1915–2001. I. Title.
BL240.3.G558 2014
215—dc23 2014012008

To Miriam

Contents

Foreword

As the newly appointed executive director of the American Scientific Affiliation, a fellowship of Christians in the sciences founded in 1941, Robert Herrmann anxiously scanned the ASA membership list, looking for potential donors for the financially challenged ASA. It was 1981 and he had just returned to the Northeast for the position with the ASA and an affiliation with Gordon College in Wenham, Massachusetts. Earlier he had had a successful career at Boston University, teaching biochemistry at the BU medical school, and was now embarking on a new phase of life.

The name "John Templeton" leapt from the roster. The renowned investor would later be known as "Sir John" after being knighted in 1987, but he was already a prominent star in the financial investment world. John Templeton had joined the

ASA as part of his broad interest in the big questions of life, particularly regarding advances in science and their connection with religion. Herrmann initiated a few phone calls with Mr. Templeton, who invited him to meet at his home in Nassau.

Their meeting at Lyford Cay in Nassau was the first of a long series of productive interactions. Sir John was brimming with ideas that he was eager to share. Herrmann was adept at listening and transcribing those ideas into prose. The collaboration resulted in several books and a life-long friendship. Herrmann helped him establish the John Templeton Foundation in 1987, while Sir John provided some funding for the ASA.

Following Sir John's death in 2008, his family recalled the great pleasure he had derived from his conversations with Robert Herrmann. To express gratitude for and acknowledgment of that relationship, the John Templeton Foundation established the annual Herrmann Lectures on Faith and Science to be organized by and held at Gordon College. The focus of the series is on the common areas of interest between Herrmann and Templeton, the big questions of science and religion.

Appropriately, the 2013 lectures by Owen Gingerich, enlarged in this book, address the question of whether there is more to physics than physics. Is there something beyond physics, that is, metaphysics, that affects our understanding of the physical world, especially as it relates to our planet and its inhabitants? Gingerich is well-positioned to pursue this big question. Born and raised in a Mennonite family, he has maintained his commitment to the Christian faith. Fascinated by astronomy from his youth, he eagerly pursued his PhD in astrophysics at Harvard University, following his undergraduate degree in chemistry at Goshen College in northern Indiana. At Harvard and the associated Smithsonian Astrophysical Observatory, his scientific career flourished where he specialized in solar and stellar phenomena. His interests gradually broadened to include history of science and he became a leading expert on Kepler and Copernicus.

Gingerich and Herrmann became acquainted through their mutual interest in science and Christian faith, occasionally sharing the stage in conferences sponsored by the ASA. Recognizing the quality of Gingerich's scientific work and its relevance to faith, Herrmann was able to arrange funding for a series of lectures on science and religion. From the late 1980s through the early 1990s, Herrmann and Gingerich traveled widely together, giving Gingerich the opportunity to share his insight into science and faith with the audiences organized by Herrmann. The ASA provided an effective network for these lectures, its members being theists active in the sciences, committed to integrity in the practice of science, and publishing the quarterly journal *Perspectives on Science and Christian Faith.*

In the 2013 Herrmann lectures, Gingerich melds his deep understanding of the physical world, the history of science, and Christian faith. He illustrates with examples from history how our understanding of the world is influenced by our cultural and religious perspectives. The magisterium of science is not self-sufficient and independent but overlaps with the magisteria of religion and culture.

Whatever the metaphysical preferences of the reader might be, the story that Gingerich weaves is a compelling one, with deep implications for addressing the big question of understanding God's planet.

RANDY ISAAC
Executive Director of the
American Scientific Affiliation

GOD'S PLANET

Prologue

ARISTOTLE HAD a word for it: *metaphysics.* It stood for his inquiry into the big questions, literally "beyond physics."

At a Renaissance university, the professor of mathematics lectured from Aristotle's *De Coelo* (*On the Heavens*) about the celestial motions of the stars and planets in their unending circles, and about the linear terrestrial motions based on Aristotle's *Physics.*

Aristotle's book *Metaphysics* was reserved for the more senior professor of philosophy. In that book the ancient Greek sage inquired into eternity and final causes. What moves without being moved? Aristotle asks. It must be the eternal heavens, with unceasing movement, he responds. But why? It must be the desire for the good, and the final cause is therefore love. In the climax to this passage,

Aristotle writes, "If then, God is always in that good state in which we sometimes are, this is wonderful, and if in a better state this is even more wonderful. And God *is* in a better state. And life also belongs to God; for actuality of reason is life, and God is that actuality; and God's self-dependent actuality is life most good and eternal. We say therefore that God is a living being, eternal, most good, so that life and continuous and eternal duration belong to God, for this is God." This, then, is the unmoved mover.

When Galileo was negotiating the terms for a position at the court of Cosimo dei Medici in Florence, he was comparatively indifferent about the amount of his salary, but he was firm about his title: "Mathematician and Philosopher to the Grand Duke." Galileo wished to discourse not only about the apparent motions of the stars and planets, but he also wanted to be credentialed to speak with authority on how the heavens were really made, the deep and even controversial issues of cosmology.

The metaphysical issues are with us even today. Included are big questions that touch on observational or experimental questions of science, but which generally lie beyond the normal bounds of science. And included are questions that fall in the domain of theology, and also rather differently, in the domain of religion. The relationship between the arena of science and the religious domain has been tense going back to the time of Galileo and beyond, but it has been particularly fraught in twentieth-century America, with issues relating to the age of the cosmos and the rise of life on earth.

And here my late Harvard colleague Stephen Jay Gould enters the fray. In 1973 Steve and Niles Eldredge, of the American Museum of Natural History in New York, published evidence that some invertebrate species evolved comparatively rapidly, followed by long periods of stasis during which the species remained constant, a pattern they referred to as punctuated equilibrium. Their critique of the standard evolutionary theory, according to which species changed slowly but continuously from small mutations, riled up the classical evolutionists, and soon Creationists seized on the controversy to declare that evolution was falling apart at the seams. Steve was appalled and began to take a greater activist role in defending evolution. In 1981 he was one of six science experts who participated in the Little Rock Creationism trial. In 1999, when he had been elected president of the American Association for the Advancement of Science, he published a defense of evolution in a volume provocatively entitled *Rocks of Ages.*

In his book Steve recounted the ongoing discord between science and religion in American society. He declared that true science and religion are not in conflict, and that each domain, or "magisterium," had valuable contributions to make provided each kept to its own territory. What he advocated was NOMA, or "non-overlapping magisteria."

NOMA sounds like a great idea, but can it, and did it ever, actually work? Can physics truly be separated from metaphysics? In Book VI of the *Metaphysics* Aristotle says it must include theology because "if the divine is present anywhere, it is present in things of this sort."

The three Herrmann Lectures that comprise this volume, given at Gordon College in Wenham, Massachusetts, in October of 2013, examine from a historical perspective how the magisteria have repeatedly overlapped over the past centuries and how today it seems unlikely that the overlaps will cease. Yet an appreciation that there are differing magisteria, with differing paths to understanding, may ameliorate the long-standing conflict.

Nicolaus Copernicus (1473–1543).

Portrait by Isaac Bullarts, *Académie des Sciences,* vol. 2, 1682.

I

Was Copernicus Right?

For many years I have puzzled about the nature of science and its theoretical structures of explanation. What gives science the ability to make predictions? In 1705 Edmond Halley predicted that a bright comet he had observed in 1682 would return again in 1758, and if it happened, he said, he hoped that candid posterity would notice that it had first been predicted by an Englishman. He was lampooned for placing the date of the comet's return well after his lifetime, so he would not have to face public scorn for such a ridiculous prognostication. But the comet did return and has borne his name ever since.

And there have been many later astronomers who envisioned planets around many distant stars, though they had little hope of actually verifying this. Today, with the recent Kepler mission, direct evidence for this prediction has been attained for nearly a thousand extrasolar planets, or exoplanets, as they are called. In another field, biologists have concluded that the ancestors of whales lived on the land, and today paleontologists have found hundreds of skeletons of early whales that still have vestiges of legs. Or in physics, we have all heard of the massive search for the so-called Higgs boson, which was predicted to exist and was finally found this past year.

This uncanny ability to make such a coherent picture of the physical and biological world has now allowed science to reign at the top of the tree of knowledge. This has not always been the case. Five centuries ago in Western civilization theology was considered the queen of the sciences, that is, the queen of knowledge. So what is the epistemological relationship between science and theology today? Are they separate magisteria, each going its own way, entirely unrelated? This is a major puzzle, and one I do not expect to resolve. Nevertheless, it is a central puzzle that I hope to address in these chapters, gradually circling around the issues from a historical perspective.

The theme of this first chapter is "Was Copernicus right?" Nicolaus Copernicus was of course the Polish astronomer who introduced the earth-shaking heliocentric cosmology. When this first chapter was presented as a lecture, I remarked that presumably nearly everyone in the audience would agree that his cosmology was right. A few might have suspected that I had a perverse reason to answer "no" and a few others may very well have hoped that I would simply say "yes, he was right" and sit down. Either response would of course have led to a scandal, and the audience wouldn't have wanted to miss that.

But then again, if Copernicus's cosmology was right, why did it take a century and a half before a majority of educated people accepted the idea that the Earth moved and the Sun stood still? So that is the particular puzzle facing us. First we should look briefly at the astronomy involved and then at the cultural and theological milieu into which it was thrown.

Let me begin by introducing Copernicus. He was a contemporary of Columbus and of Martin Luther, two other personalities who reshaped our world views. Young Nicolaus came under the patronage of his maternal uncle, who was making great strides in ecclesiastical politics, and who became bishop of Varmia, the northernmost Catholic diocese in Poland, a post in authority and power comparable to being governor of the province. Nicolaus was elected a canon of the Frombork Cathedral in Varmia, which meant that he was one of the sixteen members of the cathedral chapter, that is, its board of directors. Never ordained as a priest, he nevertheless took minor orders and had charge of one of the cathedral altars. His uncle sent him to Italy to study law and medicine, two areas of interest to the diocese. As a graduate student in Italy, Copernicus was in Bologna on the same day as Leonardo da Vinci, though it's unlikely that the eminent artist met the aspiring young canon lawyer.

Even as an undergraduate in Cracow, Copernicus had taken a keen interest in the stars and planets, and had armed himself with a few basic books in astronomy. Remember that the printing of books was not even fifty years old; had Copernicus lived a century earlier, it would have been far more

Frombork Cathedral in the northernmost Catholic diocese in Poland, where Copernicus worked as a lawyer and medical doctor. In the distance is the Vistula Lagoon, brackish waters separated from the Baltic Sea by a narrow spit of land. Photograph by Paul Gander.

difficult to obtain the sources he needed for his reform of astronomy. While in Italy he boarded at the home of Domenico Maria Novara, the university's astronomer, and while there he made some of his earliest preserved observations.

The astronomy that Copernicus studied at Cracow and then in Italy was geocentric, that is, with the Earth firmly fixed in the center of the cosmos.

Now today, if we wanted to calculate the position of Mars with a very rough approximation, we could use two circles, one for the orbit of Mars around the Sun and the other for the orbit of the Earth around the Sun, and we could connect them with some simple trigonometry. In Copernicus's day, using astronomy based on the ancient Ptolemaic system, there were also two circles, one going around the fixed Earth and the other, called an epicycle, riding on the first circle. From the point of view of geometry, the basic calculation was the same, whether you used the Sun as the center or used the Earth as your fixed reference point. The goal was identical, to know where to look for Mars as seen from the Earth. It was just mathematics, one way or the other.

But now suppose you are a student at the Jagiellonian University in Cracow in 1492, and someone came along and told you that the Sun is really a lot bigger than the Earth, and therefore the Sun, not the Earth, should be at the center of the universe, and furthermore, it didn't make any sense for the entire cosmos to spin around the Earth every day. You would no doubt have told him to get lost and to take all that nonsense with him. If

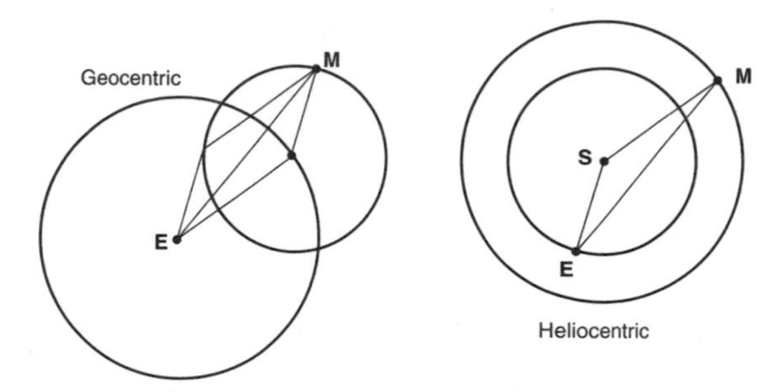

The goal of the planetary model is to find the direction to a planet in the sky, the line EM from Earth to Mars in either model. The diagram shows the simplified "kindergarten" model, without variations of speed, for example, but the important point is that the complexity is the same in either model. The triangle ESM is identical in both models. It was long known that the direction of the radius in the geocentric epicycle was always parallel to the direction to the Sun. After constructing the parallelogram, Copernicus realized that he could place the Sun at the upper-left corner of the parallelogram (as shown here) and then link all of the planets into a unified Sun-centered system.

the Earth was whizzing around the Sun, and spinning on an axis every day, a thousand miles an hour, we would surely just be spun off into space. And think how much harder it would be to walk west than to walk east! Totally ridiculous!

When I described Copernicus's heliocentric cosmology as earthshaking, the adjective was deliberately chosen, for not only did it defy common sense, but it was soon to run up against Psalm 104: "The Lord God fixed the Earth on its foundation so that it can never be shaken." For all these reasons, Copernicus was very reluctant to publish his radical ideas, although he had spent years not only working through all the mathematics, but also assembling the specific observations he needed. He was able to show that some of the parameters describing the planetary motions had changed slightly since the days of Ptolemy and some had remained the same, yet they could all be fit into a Sun-centered system. But Copernicus, as he himself put it, feared that he would be hissed off the stage.[1]

There the matter rested as Copernicus neared the age seventy, when, remarkably, a young teacher from the Protestant university in Wittenberg, Georg Joachim Rheticus, turned up on his doorstep, begging to learn what Copernicus has accomplished. Rheticus's visit, intended to last a few days, soon extended to weeks, then months, and finally to over two years. He persuaded Copernicus to publish, and eventually took the precious manuscript to Nuremberg, where it was finally printed in 1543 under the title *De revolutionibus orbium coelestium,* or *On the Revolutions of the Heavenly Spheres.* Copernicus himself received the final pages on his death bed. It was a hefty tome of 400 pages, rivaling Ptolemy's *Almagest,* the only comparable treatise.

So, what was its impact?

Let me reset the clock from Germany in 1543 to the British Isles in 1970 and invite you to join me on a school-holiday family trip from Cambridge, England, where I was on a sabbatical leave. En route to Edinburgh, Scotland, we stopped in York so I could consult with a colleague, Jerry Ravetz, who like me was on the committee to plan the forthcoming international celebrations for the five-hundredth anniversary of Copernicus's birth in 1473. And we asked ourselves that very question, what was impact of Copernicus's *De revolutionibus?*

Now, a dozen years earlier, the German-American novelist Arthur Koestler had published a history of astronomy entitled *The Sleepwalkers.* As he later confessed, he had been upset by the fact that virtually all German schoolboys knew the name of the Italian Galileo, but few could identify Johannes Kepler, the German astronomer who had discovered the elliptical form of the planetary orbits.[2] So he deliberately set out to write a book to redress the balance. As a novelist, he was prone to see the world in terms of good guys and bad guys, and thus in *The Sleepwalkers* Kepler was given the role of good guy and Copernicus and Galileo were the bad guys. In particular, he branded *De revolutionibus* as "the book nobody read" and "an all-time worst seller."[3]

On that evening in York, Ravetz and I asked ourselves who might have read Copernicus's book in the sixteenth century, and we counted fewer than a dozen names before we ran out of ideas. And then our conversation drifted off to other matters.

In Scotland two days later, while Miriam and our sons explored the Edinburgh castle, I delved into the fabulous collection of rare astronomy books at the Royal Edinburgh Observatory, and there I discovered something truly astonishing. It was not just a first edition of *De revolutionibus,* but a copy filled with marginal annotations by a reader who worked his way through the entire opus, highlighting key passages, explicating complex sections, and finding a scattering of small errors. If this book had so few readers, what was my chance that the very next copy I saw would bear the weight of heavy and perceptive reading? It just didn't add up.

With a pounding heartbeat I looked for clues to the identity of the annotator, and eventually noticed impressed into the binding the initials *E* and *R*. "Jackpot!" I thought, for these matched a name from our list of probable readers, Erasmus Reinhold, the senior professor of astronomy at Wittenberg, colleague of Rheticus, and a leading astronomical pedagogue of the sixteenth century. In my excitement I grabbed a sheet of paper and made a rubbing of the initials on the binding. *ERS* appeared on the sheet. Wait a minute! Where did that *S* come from? That let all of the air out of my balloon!

With respect to early books, I was then a mere adolescent, but I was a fast learner, and it was only a matter of days for me to find out that *ERS* was exactly what I should have expected, for in those days the town of one's birth was part of one's identity. *Erasmus Reinholdus Salveldiensis* was the Wittenberg professor's full appellation, and the mysterious *S* stood for Saalfeld, the town of his birth.

The serendipitous discovery of Reinhold's richly annotated copy of *De revolutionibus* provided a window into the way a skilled sixteenth-century astronomer looked at Copernicus's unorthodox cosmology—in fact, he essentially ignored it!—but more of that in a moment. Finding this spectacularly annotated book ignited a quest to see what the margins of other copies might contain, a possibly quixotic search to examine all possible surviving sixteenth-century examples of *De revolutionibus,* a series of globe-trotting journeys extending over thirty-five years and tens of thousands of miles. One result was a detailed census of 600 copies, and the other was a memoir ironically entitled *The Book Nobody Read,* whose last words are, "Arthur Koestler was wrong, dead wrong!"[4] In other words, every astronomer who took his occupation seriously was very likely to have had and to have read a copy of *De revolutionibus.*

Still, Arthur Koestler was a clever, educated man. How could he have been so wrong? And this is part of our puzzle, the very long time it took for a majority of educated persons to accept the heliocentric cosmology.

An overwhelming majority of us agree that it is the Earth going around the Sun and not vice versa. We know the Sun is a mass of incandescent gas, very different from the planets arrayed around it, or as Copernicus put it in his soaring cosmological chapter, "the sun, seated on a royal throne, governs the family of stars that wheel around it." Surely, Koestler must have reasoned, if scholars actually read the book, they must have promptly seen the light, and since widespread adoption of the new cosmology did not happen, it must have been a book that nobody read. It apparently never occurred to him that scholars read *De revolutionibus* as a recipe book and just did not believe that it applied to physical reality, how things really were.

Let us take the case of Erasmus Reinhold and his thoroughly annotated copy. The technical geometric passages bristle with his calculations and commentary. But the sparse cosmological chapters

Erasmus Reinhold's annotated copy of Copernicus's *De Revolutionibus,* in the Crawford Collection of the Royal Observatory, Edinburgh. Photograph by Owen Gingerich.

have scarcely any annotations. On the title page, where we might expect to encounter a comment such as "this author stops the sun and throws the earth into dizzying motion," we find instead the cryptic motto, "The axiom of astronomy: celestial motion is uniform and circular, or composed of uniform and circular parts." Here is a very different aesthetic, a golden rule for the approved technical geometry of the individual parts, not a grand unifying vision of the entire system.

And, in Copernicus's vision, it was a unified system. What he realized was that the entire entourage of planets automatically arranged themselves so that the planet with the shortest period, Mercury, orbited closest to the Sun, and lethargic Saturn, rounding the Sun in thirty years, was farthest, and all the rest fell proportionately in between. There was something irresistibly beautiful about this layout. Furthermore, this arrangement explained something that was simply a mystery in the Ptolemaic astronomy. Mars, Jupiter, and Saturn periodically stopped their eastward progress against the background stars, and moved westward for a few weeks, the so-called retrograde motion. Why did this always happen when the planet was directly opposite the Sun in the sky? This was simply a fact of nature, a "fact-in itself" as Aristotle would call it. Ptolemy couldn't explain it. But Copernicus could. With his Sun-centered plan, retrograde motion occurred as the faster-moving Earth bypassed Mars (for example), and that happened

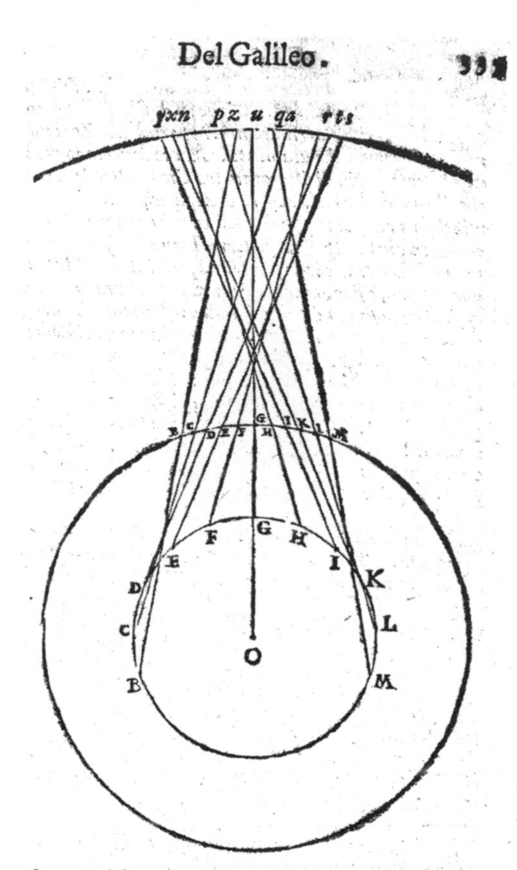

Supponendo hora, che quando la terra è in B. Gioue sia in b. ci apparirà a noi nel Zodiaco essere in p. tirando la linea retta Bbp. Intendasi hora la terra mossa da B. in c. e Gioue da b. in c. nel'istesso tempo; ci apparirà Gioue esser venuto nel Zo-
diaco

when that planet was closest to the Earth and directly opposite the Sun (see Galileo's diagram). The formerly mysterious coincidence now became a "reasoned fact." This arrangement provided a conceptual system. Copernicus had invented the solar system!

Reinhold's Wittenberg colleague, Rheticus, said that all these motions were bound together as if by a golden chain.[5] Stray, unreasoned motions now had a common cause. It was beautiful, an aesthetic vision. Rheticus even wrote an essay arguing that the new system was compatible with the Bible.

Reinhold, on the other hand, was a conservative. He was nevertheless enthusiastic about Copernicus's restoration of astronomy with his fresh observations and recalculations, and Reinhold worked diligently to bring out the volume of tables

---

Galileo's explanation of retrograde motion in the Copernican system. The faster-moving Earth (the smaller innermost circle) bypasses slower-moving Mars (the middle circle), causing Mars to appear to go backward as seen against the distant starry background (the upper arc). From Galileo's *Dialogo* (Florence, 1632), in the author's collection.

based on *De revolutionibus.* Yet the resulting *Prutenic Tables* were carefully constructed to work independently of the cosmology. They helped to popularize Copernicus's name, but a geocentrist could use the tables without feeling a heliocentric twinge.[6]

Undoubtedly Reinhold would have subscribed to the sentiment of Tycho Brahe, the great Danish astronomer, two generations later, who wrote that Copernicus's system nowhere offends the principles of mathematics, yet it throws the Earth, a lazy sluggish body unfit for motion, into a motion as swift as the ethereal torches, that is, the stars themselves.

And even Copernicus realized that his system had a serious Achilles heel. If the Earth was really whizzing around the Sun, there should be a parallactic effect on the observed positions of the stars. (You can readily observe this effect by looking at your thumb held at arm's length, first with one eye and then with the other, watching your thumb jump against the more distant background.) Copernicus had a reason why this effect was unobserved: The stars were simply too far away. "So vast, without any question, is the divine handiwork of the almighty Creator." That is precisely how he ended his soaring cosmological chapter. But in the decades that followed, serious evidence was brought forward that the stars were not all that far away, and therefore Copernicus's heliocentric cosmology could not be sustained.[7] Serious evidence, but flawed as further observations eventually demonstrated.

While the Copernican story is often told as the search for stellar parallax, I think this is a red herring. Among other things, the heliocentric cosmology was generally accepted by educated persons a century *before* a convincing stellar parallax was finally established, in 1838. Yet for a century and a half after *De revolutionibus* was published, that is, until around 1700, it was not widely accepted. Clearly something deeper was at stake. It was, I believe, the very gradual abandonment of an entrenched world view, what C. S. Lewis has perceptively described in his wonderful scholarly book, *The Discarded Image.*

The discarded image is splendidly epitomized in a late fourteenth-century Italian painting by Piero di Puccio in the Camposanto, or burial hall, that stands across the green from the Pisa cathedral and its famous leaning tower. Galileo must have walked through the hall many times when he was first a student and then an assistant professor at the university. And there, in the Camposanto, he would have seen the early Renaissance cosmos, a surviving medieval tradition, stunningly and colorfully frescoed on the wall.

Piero di Puccio's masterpiece depicted a tidy universe, the Earth in the center surrounded by the planetary spheres, and, beyond the stars, Dante's layers of the empyrean: saints and angels arrayed wing tip to wing tip, with God the Creator holding all in his arms. Puccio's universe mirrored the beliefs of popes, professors, and peasants, of merchants, monks, and mendicants. His cosmos was orderly and beautiful; "cosmetics" shares the same root. But most satisfying of all, its dimensions were comfortably human. Millions, billions, or trillions were not part of its vocabulary.

After nearly six centuries on the Camposanto wall, Puccio's fresco was shattered on July 27, 1944. A casualty of World War II, the painting was smashed to the floor in a thousand fragments. Somehow the demise of the Camposanto fresco is a fable for our times, for just as surely as that image was shattered, so was the concept it reified. Gone today is the tidy, closely bounded stage on which the great Western monotheistic religions framed their cosmologies. In its place is a vast cosmos, in which the Copernican arrangement was the opening tocsin.

Puccio's fresco was not a unique portrayal. For example, Hartmann Schedel's great illustrated coffee table book of 1493—well, maybe they didn't have coffee and coffee tables then—but the extravagantly illustrated best seller included full-page illustrations of the days of creation, ending with a magnificent view of the central Earth encircled by the spheres of the moon, the Sun, and the planets. Outermost was the ethereal sphere of heaven with the elect, all enwrapped in God's arms. In the four corners of the wood block are the four winds, quite possibly cut by a young apprentice, Albrecht Dürer.[8]

These two examples are just the tip of the iceberg as far as contemporary knowledge of the nesting of the spheres is concerned. These images depict the cross fertilization of a scientific idea with a theological picture, and this entanglement then

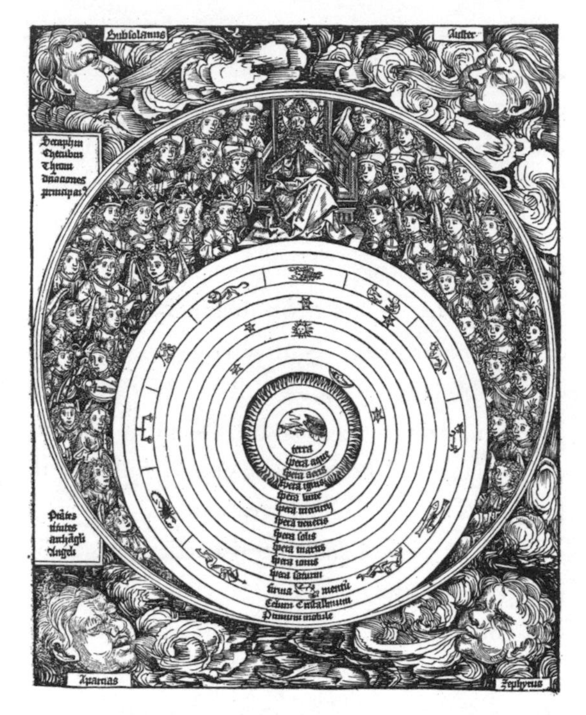

A typical fifteenth-century view of the geocentric Earth surrounded by the planetary spheres, all nestled in God's arms in heaven above. This splendid diagram appeared in Hartmann Schedel's *Nuremberg Chronicle* of 1493. The winds in the four corners are likely the work of a young apprentice, Albrecht Dürer. Author's collection.

became a very conservative force when the scientific ground began to shift. Let me now trace the scientific idea of fitting all the planetary spheres into a compact arrangement, and show how it provided a powerful background motivation for Copernicus's radical cosmology. The bottom line that will emerge in these three chapters is that science, working within it own magisterium, is far more tangled with a humanistic or theological vision than we might expect, and that the magisteria are more overlapping than we might idealistically (from a strictly scientific perspective) suppose. I now return to the historical evidence.

Nesting the planetary spheres goes all the way back Ptolemy, and he may have picked up the idea from Aristotle. During the late Middle Ages it received a further impetus in a work called *Theorica planetarum,* reputedly from Gerard of Cremona, a learned translator in Seville in the mid–twelfth century, but quite possibly from one of his mentors, John of Seville. In any event, it stemmed from Arabic sources that had preserved Ptolemy's ideas. In the middle of the fifteenth century, shortly before Copernicus was born, a Viennese professor named

Georg Peurbach updated the work under the title *Theoricae novae planetarum,* that is, the new theorica, or schemata, of the planets—not a new theory, but a clearer presentation of the nesting idea, or theorica.

The fall of Constantinople in 1453, twenty years before Copernicus was born, brought an influx of Greek-speaking churchmen into the Italian peninsula. Included was Cardinal Bessarion, who brought to Rome a Greek manuscript of Ptolemy's *Almagest,* and who was looking for a knowledgeable scholar who would learn Greek and undertake a Latin translation of this classic work. Peurbach (1423–1461) and a younger scholar also at the University of Vienna, Regiomontanus (1436–1476), accepted Bessarion's challenge. Peurbach died in his thirties, leaving Regiomontanus to complete the task. Regiomontanus, who turned out to be the greatest mathematician of his century, was a Renaissance man of great vision, and he saw a revolution at hand, coming about because of the invention of printing with movable type. Responding to the challenge, he set himself up in Nuremberg as a printer, and among other things he published the first edition of Peurbach's *Theoricae novae planetarum,* in 1472. Unfortunately he, too, died just as he was reaching the age of forty; the publication of the *Epitome of the Almagest* lagged behind, and was at last printed in 1496. Very shortly thereafter it became an essential reference source for the young Nicolaus Copernicus.

Meanwhile, Peurbach's *Theoricae novae* was reprinted repeatedly, and became a standard text for university students. In 1528 it appeared in a fine vernacular French edition, with schematic diagrams, but then in Paris in the same year it was published again (in Latin), this time in a superbly gorgeous printing with large carefully scaled diagrams.

None of this has any obvious connection to Copernicus's heliocentric idea, except that in the Regiomontanus–Peurbach *Epitome of the Almagest* there is explicitly shown the parallelogram transformation that connects the geocentric and the heliocentric geometry, something of direct use to Copernicus. But, I suspect, there is a more subtle connection to Peurbach's *Theoricae novae.*

Unfortunately we do not have any major archive of Copernicus's papers as he worked through his astronomy from geocentric to heliocentric. But we do have his autograph manuscript of *De revolutionibus* as well as an early short draft of his heliocentric system, the so-called *Commentariolus* or "little commentary," dating from around 1512. In it he lists a series of starting points, and one of them is that he detests the so-called equant, a mathematical construction adopted by Ptolemy to model the variable speed of a planet in its orbit. We know today that this is an amazingly good approximation to what is called Kepler's law of areas. What precisely was the equant; why was it needed, and why was Copernicus so put off by it?

In introducing the Copernican versus the old Ptolemaic system, I described the systems in terms of pairs of circles in motion. The revolutions in the circles, which have no beginnings and ends, could go on without end, particularly appropriate for eternal celestial motions. Popular writers often describe these as "perfect circles," which tempts me to ask what an imperfect circle is. The implication, rarely stated, is that in a perfect circle the speed of motion is constant and that the major circle is exactly centered on the Earth (in the Ptolemaic case) or on the Sun (in the Copernican cosmology). This is what I refer to as the kindergarten version. It is delightfully simple—much too simple, because it simply won't work. The ancient Babylonian astronomers and their Greek counterparts knew this full well, and their predictive schemes handled this with some clarity. The Babylonians used numerical tables, whereas the Greeks introduced geometrical models.

What the Babylonians had discovered was that the planets' motions in the sky were not uniform. Mars, for example, generally moved eastward against the starry background, but about once every two years it would come to a stop and then move westward, or retrograde, for a few months. The Greeks modeled this with the secondary circle (or epicycle), which was a good start in representing the retrograde motion (see accompanying diagram). As beginning astronomy students quickly learn, it is much easier to understand how the Ptolemaic epicycle generates the retrograde motion than how the Copernican system does, but the systems are closely equivalent, as they should be since both are attempting to reproduce the same observations.

But even this is not enough to make a good approximation of the planetary motions. As the

Babylonian observers soon found, when they averaged out the retrograde motions, Mars moved much faster when it was moving through the constellation Cancer than when it was on the opposite side of the sky, moving through the constellation Capricorn. Furthermore, the Babylonian observers discovered that the length of the retrograde loops depended on where in the sky they took place. For example, when the retrogression took place in Cancer, it covered a longer arc of the sky than when it happened in Capricorn. The Babylonians approximated this by making arithmetic tables for the different parts of the zodiac.

The Greeks, however, were into geometry, and around 125 A.D. Claudius Ptolemy constructed his epicyclic theory, in which the motion of the secondary circle, or epicycle, generated the retrograde motion. Suppose that the epicycle moves uniformly around the large carrying circle, but that the Earth is about 10 percent off-center (which is close to the case for Mars). When the Earth is closer to the Martian epicycle, the planet will appear to move faster. This is a good start to representing the motions of Mars, but alas! The lengths of the retrograde arcs

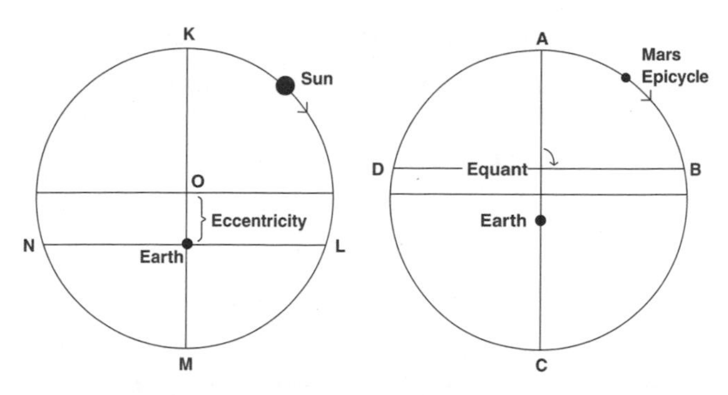

In the eccentric model that Ptolemy used for the Sun (*above, left*), the Sun moved uniformly around the center point O. The quadrants KL, LM, MN, and NK each contain 90° as seen from Earth. Because of the eccentricity, the Sun appears to move more quickly through the Earth-centered quadrant LM than through quadrant NK. If Ptolemy could have measured precisely the apparent diameter of the Sun, he would have found that his eccentricity of the Sun's orbit was twice what it should be.

In the equant model Ptolemy used for the planets (*above, right*), the equant acts like the hand of a clock, driving the planet's epicycle uniformly through each quadrant as seen from the equant. As seen from the Earth, however, the speed in the longer arc BC must be faster than in the shorter arc AB because the total time in each arc is the same.

come out all wrong. Something else was needed, and here Ptolemy was devilishly clever, though he had to ignore Aristotle's theoretical instructions that the circles should have uniform motions, that is, unvarying speeds. Aristotle just was not clued in to the detailed motions of the planets.

What Ptolemy did was to establish another seat of uniform motion, called the equant. As may be seen in the diagram, it is placed near the middle of the carrying circle opposite the Earth and at the same distance from the center. Amazingly, Ptolemy's invention of the equant makes not only the length of the retrograde loops come out just right, but also the varying speeds of the planets around the sky work out as observed. So why did Copernicus detest the equant?

I have been puzzled by this obsession for several decades, ever since I began to look at Copernicus's mathematical mechanisms in some detail. The answer, I now believe, was lying in plain sight all along, like the purloined letter. Let us look closely at Copernicus's diagram of his heliocentric system, here shown in his working manuscript of *De revolutionibus.* The Sun is at the center and the

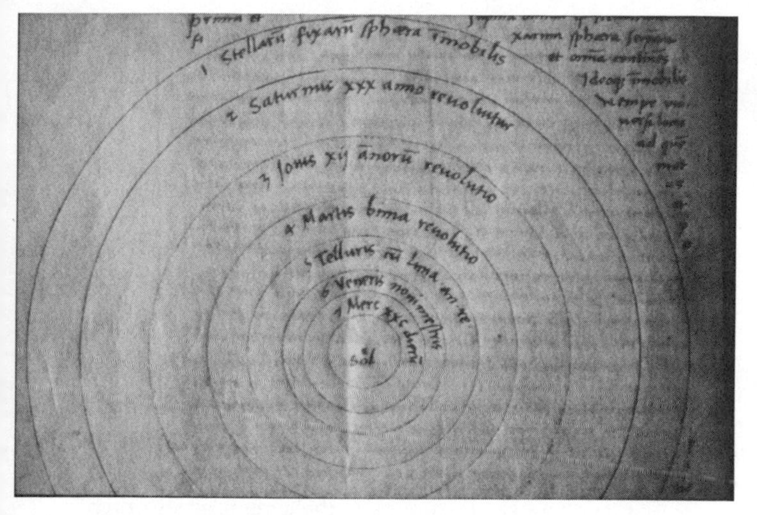

Copernicus's autograph manuscript for *De revolutionibus,* preserved in the Jagiellonian Library at Cracow University, shows the heliocentric cosmos. Note that Copernicus has included the period of revolution for each planet, emphasizing the "sure connection between the time of revolution and the distance from the Sun." Photograph from the author, July 1972.

planets are ranked around the Sun in the order of their periods of revolution: Mercury, Venus, the Earth with its moon, Mars, Jupiter, and Saturn. An equant for each of these planets would require another sphere, whose center in the case of Mars,

Jupiter, or Saturn would lie on top of the sphere for Mercury. This is the first serious attempt to arrange the planets into a system, the invention of the solar system. But with an equant sphere for each planet, what a confused mess that would be! Something to be avoided at all cost! And this is precisely what Copernicus manages to do.

This diagram does not show orbits. Instead, like Peurbach's *Theorica,* it shows zones—zones that can easily accommodate a subsidiary epicyclet for each planet, a little circle that neatly substitutes for the disdained equant. Such an alternate geometry that exactly duplicates the equant was worked out by Islamic astronomers in the fourteenth century, but no satisfactory pathway of transmission to Europe has been found, so it is highly controversial whether Copernicus reinvented it himself or if he had got wind of the Arabic invention, possibly during his graduate studies in Italy.

That Copernicus's heliocentric diagram shows zones becomes particularly apparent when we notice that the Earth with its moon—*terra et luna*—is included in the third zone. Here was now a compelling harmonic arrangement for those with eyes to see it, everything tidy and not spilling outside its own zone. This was the deeply aesthetic judgment that brought Copernicus to his radical cosmology with its well-stacked harmonic arrangement.

Now traditionally there has been some prejudice against accepting a theory on the grounds of its aesthetic appeal. And such an appeal was to a considerable extent in the eyes of the beholder. A pious Lutheran or Catholic could have felt that nothing could be more blessed and beautiful than a closed universe surrounded by the ethereal heaven with God the Father, the angels, and the elect. Besides, it was ridiculous to think of the Earth spinning at a thousand miles an hour. However, to Galileo, it was not clear that the motion could be felt, and in any event, he stated that he could not admire enough those who accepted Copernicus's ideas *despite* the evidence of their senses.[9] And then there had been the Englishman, Thomas Digges, who in his perpetual almanac for 1576 (see diagram) had included an emporium for the elect among the heavens, a compromise between Copernicus's vast universe and a place for the hereafter. It was the first published diagram to show the stars extending beyond a thin sphere.

43

A perfit defcription of the Cœleftiall Orbes, *according to the moft auncient doctrine of the Pythagoreans. &c.*

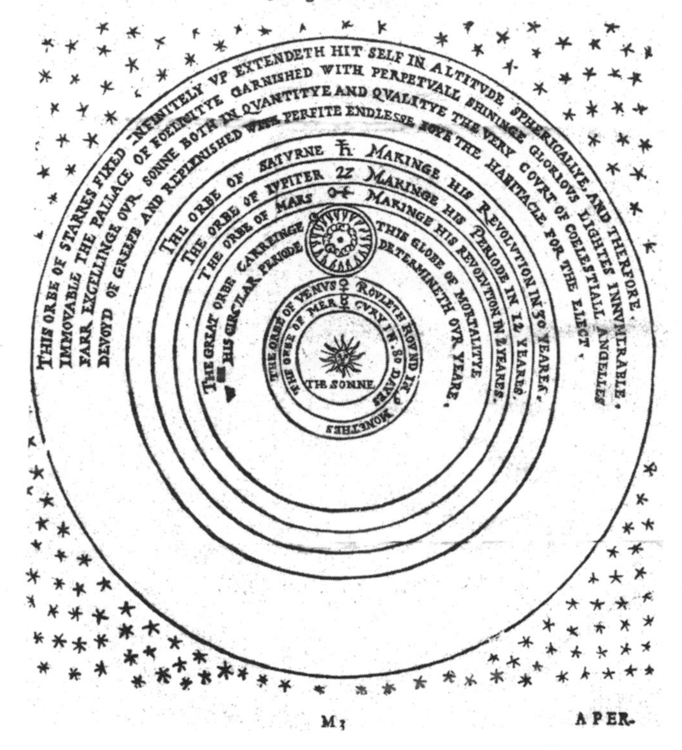

Thomas Digges's heliocentric system, originally published in 1576, showed for the first time the stars distributed through space rather than being contained in a thin shell centered on the Sun. This is from the unique surviving diagram from the edition of 1592. Author's collection.

All of these ideas tiptoed around the circumstance that as the seventeenth century began there was no physical evidence in favor of the heliocentric system beyond its unifying aesthetic. Arthur Koestler's notion that simply no one had read Copernicus's book was scarcely a credible explanation for the failure of widespread adoption of a heliocentric cosmos. Now many of the most important breakthroughs in science have been major unifications: terrestrial and celestial physics in the case of Newton, electricity and magnetism by Maxwell, space and time by Einstein, and so on. What was holding up the adoption of the wonderfully unified planetary system of Copernicus?

It is fascinating to examine the opinion of Tycho Brahe, a true giant among observational astronomers of all time. He knew that Copernicus, half a dozen decades earlier, had made a powerful unifying step, but nearly everyone supposed it was just a hypothetical construction. Yet, standing under the stars with his amazing instruments night after night, he must surely have decided that he was looking at something real, and not just imaginary constructs. But it seemed unreasonable for this heavy, sluggish Earth to be moving so swiftly, and furthermore it was surely un-Biblical because of Psalm 104 and similar passages that extolled the stability of the Earth. So whenever he condemned the motion of the Earth, he always listed physics first and the Bible second. The Bible! By our standards, that's in a completely different magisterium. Tycho shows us how overlapping the magisteria can be.

Today nearly every introductory astronomy textbook carries two proofs of the motion of the Earth. The rotation of the Earth on its axis is demonstrated by the changing plane of swing of the famous pendulum first swung publically by Leon Foucault on a night in February 1851. The revolution of the Earth around the Sun is verified by the annual parallactic motion of nearby stars, already mentioned by Copernicus, but not convincingly measured until 1838. But these two proofs were not available until the nineteenth century.

It was Cardinal Roberto Bellarmine in Rome, the leading theologian of his day, who challenged Galileo to find an apodictic proof of the motion of the Earth. The cardinal made very clear that he was unwilling to concede the motion of the Earth in the absence of such an infallible proof when he added, "If there were a true demonstration, then it would be necessary to be very careful in explaining Scriptures that seemed contrary, but I do not think there is any such demonstration, since none has been shown to me. To demonstrate that the appearances are saved by assuming that the Sun is at the center is not the same thing as to demonstrate that *in fact* the Sun is in the center and the Earth in the heavens."[10] One wonders how Bellarmine would have responded to these modern proofs.

Suppose that the Foucault pendulum had been set in motion with its shifting orientation of the swing. What would Bellarmine have made of that? Well, why not suppose that the influences of the stars whirling around the Earth caused the plane of oscillation of the pendulum to rotate? This is not a frivolous way out, for it is the general relativistic explanation. And what if the annual stellar parallax had been found? Well, why not let each star have its own tiny epicycle, cycling around each year? I think such an explanation would have naturally occurred to Bellarmine. You may immediately think of Ockham's razor, that the simpler explanation would surely prevail. But remember that Ockham's razor is not a law of physics. It is an element of rhetoric, in the tool kit of persuasion. In the absence of new physics, myriad epicycles might not have been an obstacle to keeping the Earth safely fixed.

Of course this is not entirely fair to Bellarmine, because since the time of Galileo the background has entirely changed. I am put in mind of a totally different situation, namely, the change of opinion expressed in two very famous rulings of the American Supreme Court, on the question of racial segregation.[11] The first of these was the 1896 case of *Plessy v. Ferguson.* In 1890 the Louisiana legislature passed a law "providing for separate railway carriages for the white and colored races." The law, which required that all passenger railways provide separate cars for blacks and whites, stipulated that the cars be equal in facilities and banned blacks from sitting in white cars and whites in black cars.

In 1892 Homer Plessy, who was seven-eighths white and one-eighth black, purchased a first-class ticket for a trip from New Orleans and took a vacant seat in a whites-only car. It was an orchestrated case to challenge the law. Duly arrested and imprisoned, Plessy was convicted in a New Orleans court of violating the 1890 statute. He then filed a petition at the Louisiana Supreme Court against the trial judge, John H. Ferguson, and the case, after being lost at the state level, was in 1896 appealed to the Supreme Court, where, as *Plessy v. Ferguson,* it again lost. The Court ruled that, while the object of the Fourteenth Amendment was to create "absolute equality of the two races before the law," such equality extended only so far as political and civil rights (such as voting and serving on juries), not "social rights" (for example, sitting in a railway car one chooses). As Justice Henry Brown's opinion put it, "if one race be inferior to the other socially, the constitution of the United States cannot put them upon the same plane."

Thus the notion of "separate but equal" became enshrined in U.S. law for over half a century, during which time it became increasingly apparent that many schools (for example) were separate but demonstrably unequal. Then, in 1954, another landmark decision by the Supreme Court abruptly overturned the "separate but equal" doctrine. In

Brown v. the Topeka Board of Education, the Warren court ruled unanimously that segregating children by race in public schools was "inherently unequal" and violated the Fourteenth Amendment.

In making their decisions, the judges are in principle supposed to rely on the briefs submitted on each side of the argument rather than outside reading or discussions. In most respects the issues of law were the same in *Brown v. Board of Education* as in *Plessy v. Ferguson.* What had changed? Essentially the social environment and public ideals of what was right. It was a slow change, and still far from unanimous. And the judges of the Warren court could not help being influenced by the evolution that had taken place in public opinion.

Likewise, throughout the seventeenth century, the public acceptance of the Copernican cosmology was slow, far from unanimous, and based not on proofs but on the persuasion of what was increasingly seen as a coherent system. A very important element in its acceptance was Galileo's *Dialogue on the Two Chief World Systems,* published in 1632. Although he had no solid proofs of the motion of the Earth, Galileo argued persuasively, making it intellectually respectable to believe in the Copernican system. I like to say that it was the book that won the war.

Galileo believed that he had two particularly persuasive arguments showing that Copernicus was right. One had to do with the rotation of the Sun, which he had deduced from the motion of the sunspots that he had observed. It is, frankly, a fallacious argument, though probably convincing to the majority of his readers.[12] The other was the rhythm of the tides, which he thought could only be explained by the motion of the Earth, and he even made a snide remark against Kepler for superstitiously believing in the moon's dominion over the tides. Pope Urban VIII, who had apparently given a nod of approval to Galileo's writing a cosmological book, vetoed the idea of titling the book "On the Flux and Reflux of the Sea," because that would give too much emphasis to what Galileo considered a physical proof of the Earth's motion. "After all," Urban said, "God in his infinite power and knowledge could have created the tides in many other ways including some beyond human intellect."[13]

Subsequently Galileo made a stupid misjudgment: in his book he put the Pope's argument in the mouth of one of his characters, Simplicius, which happened to be the name of a sixth-century Aristotelian commentator, but which his readers noticed as a pun on "simpleton." This was undoubtedly a principal reason Urban stopped talking with Galileo and put him under house arrest for the rest of his life.

There was a significant reason Galileo's *Dialogo* did not entirely win the day. Note its full title, *Dialogue on the Two Chief World Systems.* Galileo deliberately avoided another world system, one that had been proposed in 1588 by the Danish astronomer Tycho Brahe. In the Tychonic system the Earth was solidly fixed in the center of the universe and the Sun revolved around it, while the planets were carried in thus orbits around the moving Sun. You can see the Tychonic arrangement outweighing the Copernican system in the frontispiece of the Jesuit Riccioli's *Almagestum novum.* Giambattista Riccioli's hefty two-volume set included a strong argument against the motion of the Earth. With the comparatively poor telescopes used by him and his

This frontispiece from Jesuit Giambattista Riccioli's *Almagestum novum* (1651) weighs the heavier geo-heliocentric system against the lightweight Copernican system, while the discarded Ptolemaic system lies below with Ptolemy saying, "I am raised up by being corrected." The scripture passages at the top are from Psalms 19:2, 8:3, and 104:5. Author's collection.

contemporaries, it seemed that stars appeared not as pure points of light, but apparently as magnified dots. If the stars were at the vast distances required by Copernicus for them to not show an annual parallax, they had to be ridiculously huge. And a powerful group of educators in the Catholic countries, the Jesuits, promulgated Tychonic geocentrism.

In 1674 the English polymath Robert Hooke, who would become secretary of the Royal Society in London, summarized the state of play of the arguments. The problem of the Earth's mobility, he wrote, "hath much exercised the Wits of our best modern Astronomers and Philosophers, amongst which notwithstanding there hath not been any one who hath found out a certain manifestation either of the one or the other Doctrine."[14] Thus, he suggested, people let their prejudices reign. Some "have been instructed in the Ptolemaick or Tichonick System, and by the Authority of their Tutors, over-awed into a belief, if not a veneration thereof: Whence for the most part such persons will not indure to hear Arguments against it, and if they do, 'tis only to find Answers to confute them."[15]

Hooke then confirms what I have been arguing, namely that the best and most persuasive reason for adopting the Copernican system up through his time was the proportion and harmony of the world: "On the other side, some out of a contradicting nature to their Tutors; others, by as great a prejudice of institution; and some few others upon better reasoned grounds, from the proportion and harmony of the World, cannot but embrace the Copernican Arguments."

But, Hooke allows, "what way of demonstration have we that the frame and constitution of the World is so harmonious according to our notion of its harmony, as we suppose? Is there not a possibility that things may be otherwise? nay, is there not something of a probability? May not the Sun move as Ticho supposes, and that the Planets make their Revolutions about it whilst the Earth stands still, and by its magnetism attracts the Sun and so keeps him moving about it?" There is needed, Hooke declares, an *experimentum crucis* or "crucial experiment" to decide between the Copernican and Tychonic systems, and this he proposed to do with a careful measurement of the annual stellar parallax. I will not describe Hooke's attempt, but let me merely state that Hooke thought he had confirmed the effect and therefore the Copernican arrangement. Unfortunately for him, the effect was much more subtle than he imagined or measured.

Nevertheless, by the end of the seventeenth century, the tide had begun to turn toward Copernicus. What had happened between Hooke's attempted *experimentum crucis* in 1674 and the century's end in 1700? Isaac Newton's *Philosophiae naturalis principia mathematica* in 1687 described a solar system under gravitation around a Sun vastly more massive than any of the planets, yet holding them in orbit by the mysterious but mathematically expressed gravitational power. Even the wayward comets fell into elliptical orbits rounding the Sun. Likewise, calculations showed that people would not be spun off into space by the rotation of the Earth. Here was an awesome coherency, persuasion par excellence. Copernicus was indeed on the right track.

One prediction made by Newton in his *Principia* was that the Earth, because of its rotation, should bulge out at equator, but at the time of its publication, the geodetic measurements of the Earth were not precise enough to establish this. Not until 1736, with the results from the arctic expedition of Maupertuis, was Newton's prediction confirmed, and Maupertuis became known as the man who flattened the Earth (see accompanying figure). As for the motion of the Earth around the Sun, a phenomenon known as aberration was discovered by James Bradley and announced in 1729. But the long sought and subtle stellar parallax, the periodic change in the positions of nearby stars, remained undetected until 1838, when it was found by the German astronomer Friedrich Bessel.

So, you may have thought it was completely absurd when I asked the curious question, "Was Copernicus Right?" But I hope we can now realize that it is not as bizarre as it might have first appeared, because for a century and a half the question was not settled. And we can also see that cultural attitudes, especially including religious beliefs, play a significant role in what passes as a proper scientific understanding. Did the problem

In an expedition to Lapland, Pierre-Louis Moreau de Maupertuis (1698–1759) measured the shape of our planet to confirm Newton's predictions. He appears here in his Lapp costume as he flattens the Earth. Author's collection.

of overlapping magisteria fade away when in 1838 the Catholic Inquisition finally took Copernicus's book off the *Index of Prohibited Books?* We will have another look at this problem when we ask, "Was Darwin Right?"

Charles Darwin (1809–1882). Lithograph by
Thomas Herbert Maguire, 1849.
Courtesy of Wellcome Library, London.

2

Was Darwin Right?

WHILE VIRTUALLY everyone agrees that Copernicus was right, a substantial minority, particularly in America, is convinced that Charles Darwin and his epoch-making book, *On the Origin of Species,* is *not* right. A recent Gallup poll shows that only four out of ten Americans believe in evolution, and for frequent churchgoers, the number drops to one out of four.[1]

There is a wide spectrum of reasons why people are reluctant to accept evolution. At one end are those who have difficulty reconciling the Genesis creation story with the long, slow eons of the evolution picture. Others perhaps are distressed by the role of random chance versus God's proactive hand at work. Some may rebel at the idea that we share the same family tree with chimps and gorillas. And more sophisticated critics of evolution sometimes declare that the creation of the tree of life is a nonrepeatable event and therefore cannot qualify as a science, which leads to a discussion of the nature of science itself.

As for the nature of science, I cherish an insight I got from the late Philip Morrison, an institute professor at MIT, who declared somewhat paradoxically that science is a verb, not a noun. By that he meant that science was an activity, a means of inquiry, and not a collection of subjects like astronomy, botany, chemistry, paleontology, and so on.

This distinction came to the fore a few decades ago when the Harvard faculty undertook to update their general education program, which had been put in place shortly after World War II. Students typically took four courses through each year, or sixteen in all in their four years of undergraduate study, and each student was obliged to take at least one science course during that time. But over the years the program had become diluted, so students could fulfil the science requirement, for example, by taking a course in medical ethics. The physicists were disturbed by the fact that while many students had an entire year of physical science, some had none whatsoever, and as a compromise agreed that every student should have a minimum of a half course in physical science paired with a half course in biological science.

Now at that time my friend Stephen Jay Gould had been giving a very successful general education course in evolutionary geology. The physicists were sure that was not a course in physical science; the biologists also doubted that it was a proper course in biological science. This promised to leave Gould's class in a curricular limbo. So Steve and I got together to brainstorm about the varied nature of science. Some sciences are strongly classificatory or historical in nature, and others are undergirded by natural laws from which conclusions can be deduced and matched with experimental results. We proposed that all the students should take a half course in each kind of science. The separation was a little fuzzy, but clear enough to include evolutionary geology, so Steve's course found a niche in the curriculum.

Although evolutionary geology depends strongly on classification and on reconstructing the historical record of rocks and fossils, important aspects of it can be framed as hypotheses and tested by observational data. Incidentally, the same is very much true of a great deal of astronomy, which is also an observational science. (It's not easy to bring a galaxy or a supernova into the laboratory, for example.)

So let us place Darwin's work in a historical context. In the early decades of the nineteenth century the foundations of modern geology were just being laid. In particular, geologists (often amateurs) were beginning to trace the strata soon identified with geological periods long past. Included were the Carboniferous, associated with the coal measures in the British Midlands, and the Devonian, associated with Devonshire and the White Cliffs of Dover. Lower and hence earlier strata found in Wales were named Cambrian after an early Celtic name for Wales, and somewhat more recent strata in West Anglia were named Silurian for the Silures, an ancient tribe from that area. In Russia there were extensive strata more recent than the English coal measures, and these were named Permian, after Perm, an old Russian town. These strata, stacked in order, gave rise to what is called the geological column. In the United States, British geologists found matching strata, so it was gradually realized that the geological column was a worldwide phenomenon.

As the strata were being delineated, geologists began to understand that particular fossil species were often associated with particular geological strata, and furthermore, the simpler life forms—trilobites, for example—were concentrated in the oldest strata, and more complex creatures such as sharks occurred only in newer layers.

This was the background against which the young Charles Darwin began his career as a naturalist, and it was the puzzle he would wrestle with in his *On the Origin of Species.* How could the progressive development of life species observed in the geological column be accounted for? This was not the problem he set out to solve when, in 1831 as a fresh Cambridge University graduate, he boarded the H.M.S. *Beagle* for a five-year round-the-world trip. He was not brought on board as a naturalist—that was reserved for someone else—but he was there to provide dinner conversation for Robert FitzRoy, the captain and great-grandson of King Charles II, to keep him from going nuts on a long lonely voyage during which naval protocol prevented him from dining with the junior officers. That this role was essential is reinforced by the fact that in 1865, long after his cruise on the *Beagle,* FitzRoy did commit suicide.

Besides being a worthy dinner companion, Darwin proved to be a very astute observer and collector of natural history specimens. What first struck him, when the ship reached the equatorial regions of Brazil, was the overwhelming, lush verdancy of tropical vegetation, something almost unimaginable for a young man born and bred in temperate England. And then, on the Argentinian pampas, he found astonishing fossils of giant and extinct sloths, which seemed somehow related to the much smaller armadillos still living in those grasslands. Could there be a connection between the fossil and the living species? As he wrote much later, after his return and after discussions with the geologists in London, "This wonderful relationship in the same continent between the dead and the living, will, I do not doubt, hereafter throw more light on the appearance of organic beings on our earth, and their disappearance from it, than any other class of facts."[2]

After mapping harbors along the Argentinian coast, Captain FitzRoy took the *Beagle* down toward the southernmost tip of South America. There Darwin was astonished by the "miserable, naked, cannibals" in Tierra del Fuego, and by the immense gulf separating the natives from his English peers. "What a scale of improvement is comprehended between the faculties of a Fuegian savage and a Sir Isaac Newton!" Darwin exclaimed.[3] To his mentor back in Cambridge, he wrote, "The Fuegians are in a more miserable state of barbarism than I had expected ever to have seen a human being. In this inclement country, they are absolutely naked, & their temporary houses are like what children make in summer, with boughs of trees. I do not think any spectacle can be more interesting, than the first sight of Man in his primitive wildness."[4] Nevertheless, he recognized them as being of the same species as himself, something that would in time evoke disbelief from some of his critics.

Eventually the *Beagle* sailed into the Pacific and up the Chilean coast just as an unexpected phenomenon occurred, a mighty earthquake that destroyed the town of Concepcion and raised part of the shoreline by nine or ten feet. It did not take Darwin long, as a geologist, to conclude that over long periods the Andes themselves could be raised by repeated earthquakes. As he put it, "The Earthquake and Volcano are parts of one of the greatest phenomena to which this world is subject."[5] To his sister Caroline, Darwin wrote, "I was very glad we happened to call at Concepcion so shortly afterwards: it is one of the three most interesting spectacles I have beheld since leaving England—A Fuegian Savage; Tropical Vegetation; & the ruins of Concepcion."[6]

Still awaiting Darwin was the most significant stop of the entire journey, the Galapagos Islands. Although Darwin had expected geology to be the main interest of the islands, it turned out to be the fauna and flora that would prove to be so seminal to his thinking about the origin of species. Darwin's earlier introduction to the verdant lushness of the tropical Brazilian forests in no way prepared him for the comparative sterility and equatorial heat of the Galapagos. The few trees

Iguana reptiles, Fernandino Island, Galapagos. Darwin described these lizards as having "a singularly stupid appearance." Photograph by Owen Gingerich.

were stunted, and there were no native mammals. It was, however, a reptilian paradise (see accompanying figure). Huge aquatic lizards swarmed the beaches, and the few species of terrestrial iguanas seemed to be everywhere. To Darwin, these creatures were clumsy, hideous, and disgusting. There was an abundance of giant tortoises, some so large that six men were required to lift them. And there were birds so tame they would land on the shoulders of the men, having no fear of a creature hitherto unknown to them.

Yet his reconnaissance of the Galapagos Islands proved to be a remarkably lost opportunity. Near the end of their visit to the islands, FitzRoy and Darwin met the Englishman serving as governor of the islands, and from him Darwin learned an interesting fact: the governor maintained that he could tell at once, by observing the shape of the shell, from which island any tortoise was brought. Unfortunately the specimens that went home in the *Beagle* were too small to confirm this, but the claim echoed in Darwin's mind much later as he began to appreciate the differences in the flora and fauna on the various islands. As he then wrote, somewhat ruefully, "It never occurred to me, that the production of islands only a few miles apart, and under the same physical conditions, would be dissimilar. I therefore did not attempt to make a series of specimens from the separate islands."[7] Elsewhere he wrote, "It is the fate of every voyager, when he has just discovered what object in any place is more worthy of his attention, to be hurried from it."[8]

There still remained the long journey across the Pacific, over the Indian Ocean, around the Cape of

Good Hope, and northward through the Atlantic. In Bahia, Brazil (where the ship had detoured to make a final longitude determination), a homesick and probably seasick Darwin wrote his last letter from the ship, to his sister Susan. "This zig-zag manner of proceeding is very grievous. . . . I loathe, I abhor the sea, & all ships which sail on it."[9] The *Beagle* left Brazil on August 12, 1836, still 5,000 miles from home. The ship anchored at Falmouth on October 2, and that night the twenty-seven-year-old Darwin was on the coach for his home in Shrewsbury. The voyage had lasted four and three-quarter years.

When Darwin returned from the voyage of the *Beagle,* his mentor in Cambridge advised him that it would take twice as long to describe and classify his specimens as it took to collect them. The young naturalist thought this prediction was preposterous, but in the end it required essentially a decade for Darwin to complete the publication of his data and his geological reflections. Captain FitzRoy urged him to publish his journal of the voyage, and in the first instance he did so in 1839 as an appendix to FitzRoy's official account of the journey. When a major London publisher saw it, he realized that the account would make a great popular travelogue, so he bought the copyright. Darwin's *Journal of the Voyage of H.M.S.* Beagle *Round the World* has been in print continuously ever since.

While Darwin had been voyaging, it had become increasingly clear from the fossil evidence that the older the layers, the more primitive the life forms. The geological column was a book showing the history of the Earth and of the creatures upon it. How these wonderfully diverse creatures originated and how they were related was the mystery that challenged him. The eminent Victorian scientist John Herschel labeled the process by which extinct species were replaced by new ones the "mystery of mysteries," and Darwin duly noted Herschel's phrase in one of his little notebooks.

The vast majority of naturalists at this time rejected the notion that species were transmuted from one form to another. Their work was built firmly on the assumption that each organic form was miraculously created by the Deity, though many of them allowed that the forms progressed through the geological ages. Darwin, however, was keenly aware that within a given species there was a certain amount of variation, and he knew that through selective breeding pigeon fanciers had been able to produce several lines of unusual pigeons. Perhaps, given enough time, one species could transmute into another. If so, this would afford an explanation for two geographical-distribution puzzles Darwin had encountered during his *Beagle* voyage. One concerned the armadillos of South America. The bony sheathed *Glyptodon*, whose fossil bones he had collected, had seemingly been replaced by several species of armadillos not found elsewhere, apparently transmuted descendants of the earlier monster. The other was the distribution of species on the Galapagos Islands, so similar to that on the western side of South America, yet subtly different. It was as if a few pioneer animals had managed to get to islands, and subsequently evolved into various related species.

Then, almost fortuitously, in October 1838 Darwin read the essay by the late Anglican clergyman and economist, Thomas Robert Malthus, on the problem of ever-increasing world population. Malthus argued that population, when unchecked,

Skeleton of a *Glyptodon,* a large armored relative of armadillos, a now-extinct mammal that lived during the past two million years, on exhibit at Harvard's Museum of Natural History. Photograph by Owen Gingerich.

would result in eventual famine, and the excess organisms would starve. Now inspiration struck. It was only a short step from this to his theory of natural selection. Population was necessarily limited by whether it had something to eat. Natural selection would take place, with the most fit individuals more likely to survive in the competition for food. In his later autobiography Darwin wrote,

"I happened to read for amusement Malthus on *Population* [and] it at once struck me that under these circumstances favourable variations would tend to be preserved, and unfavourable ones to be destroyed. The result of this would be the formation of new species."[10] By 1844 Darwin had his basic ideas in hand, but he wanted more examples to buttress his theory, and he was reticent to challenge the common belief that God Himself had individually created the species by direct fiat.

Darwin shared his ideas with a few confidants, including Harvard's distinguished botanist Asa Gray, but he still remained reluctant to publish. Gray was one of the few persons to whom Darwin disclosed his theory before the publication of *On the Origin of Species.* After two years of correspondence, Darwin had gradually appreciated that Gray could be trusted with his radical ideas. "But as an honest man," Darwin wrote in the summer of 1857, "I must tell you that I have come to the heterodox conclusion that there are no such things as independently created species—that species are only strongly defined varieties. I know that will make you despise me."[11]

Gray, however, recognized the significance of Darwin's scientific contribution, and he took this revelation in stride. As an orthodox Christian, a Presbyterian firmly devoted to the faith expressed in the Nicene Creed, Gray saw Darwin's theory as evidence for the way God worked in nature.[12]

In September 1857, Darwin sent Gray a précis of his theory of natural selection.[13] The following summer Darwin was taken by surprise when a similar thesis arrived from the naturalist Alfred Russel Wallace, who was collecting data in the Malay Archipelago. Darwin had been working on his theory for at least fifteen years, slowly gathering supporting evidence, but he had not published anything about it. The synopsis Darwin had sent to Gray now became particularly important for establishing his independence and for having formulated his theory first. Urged into action by the appearance of Wallace's essay, Darwin at last concentrated on writing an introduction to his theory. In scarcely a year, his *Origin of Species* was published, on November 24, 1859, and the first edition of 1,250 copies was sold out within a day. Its title was *On the Origin of Species by Means of Natural Selection, or the*

Preservation of Favoured Races in the Struggle for Life. Others had proposed transmutation of species, but Darwin and Wallace were the first to propose a mechanism. "Survival of the fittest" eventually became the slogan for the first pillar of Darwin's theory.

The second pillar of his theory, that one species could transmute into another, Darwin would call "descent with modification." By implication, Darwin meant that every creature had a mother. This seems pretty reasonable to us today, that flies are not born of garbage and eels do not just emerge from mud with no ancestors. But at that time so-called spontaneous generation was considered a realistic explanation, and it was not until 1859 (the same year that *On the Origin of Species* was published) that Louis Pasteur in Paris refuted this notion with a brilliant series of experiments, though Pasteur's work still remained controversial for several decades.[14]

"Survival of the fittest" did not rest well with much of English society. Was the world really created so that less fit children would starve? Could God have made such a cruel world? This, however, is a problem of theodicy, the problem of evil, and the theory of evolution is a solution to the origin of species, not itself a cause of cruelty. Suffering would not go away if the theory of evolution were suppressed. After all, it is a clear fact of nature that most if not all species produce more offspring than required simply to replace their parents.

What was more troubling to the biologists was how variations could be maintained to allow transmutation to occur. Darwin had little notion of how variations arose. But variations there were, and a typical question was, how could the variations in pigeons get preserved in subsequent generations without a pigeon fancier at work? At that time it was supposed that inherited characteristics were a blending of qualities from each parent. But an unusual trait would become diluted in subsequent generations as it was blended with more normal traits. It seemed that Darwin's idea would not work, and by the end of the nineteenth century, the pillars of his theory were pretty much in eclipse.

The key to the early twentieth century revival of evolution was the recovery of the overlooked work of Gregor Mendel on genetics, originally published in 1866, which had remained unknown to Darwin and most of his contemporaries. Mendel had worked as a systematic experimental botanist, exploring how different traits in sweet peas were transmitted in a mathematically predictable way. His work led to the idea of genes, and because the genes came on paired chromosomes in living cells, there could be a dominant gene and a backup, or recessive, gene. Rather than qualities being blended from each parent, they moved in block fashion from one parent or the other. If there was a change in a gene, called a mutation, it could be passed on over the generations rather than blended out of existence.

The new understanding of genetics revitalized the theoretical study of natural selection in the early decades of the twentieth century; at the same time, conspicuous evidence for the evolution of the species came from the enormous enhancement of the fossil record in the century that followed the *Origin*. For example, dinosaurs were essentially unknown to Darwin when he wrote his book. In the years that followed, two rival bone hunters, Edward

Cope from Philadelphia and Othniel Marsh from New Haven, worked the fossil beds in the American West to reveal not only a huge reptilian fauna, previously undreamed of, such as monster dinosaurs, but also hundreds of extinct mammalian species, including one of the most complete sequences, that of fossil horses. The early Spanish conquistadores may have thought horses were their gift to America, but in truth the horse was the American continent's gift to the world. From the point of view of the modern horse, the fossil record shows a linear progression from a fox-sized, multitoed mammal up to the hoofed *Equus* of today, but it also shows dead-ended side branches as well as times when as many as a dozen different horse species simultaneously roamed the continental savannahs. And, during the twentieth century, great strides were made in actually pinning dates onto the layers of the geological column, something made possible by analysis of radioactive isotopes in the rocks. Again, these were developments Darwin could hardly have imagined.

The increase of the fossil record continues apace today, and several discoveries deserve special mention because they were predicted, and paleontologists went to specific localities where the layers were of the proper age and there managed to find the previously unknown fossil species, what might be termed "missing links." One concerns the origin of whales from land creatures related to the hippopotamus. When semiaquatic mammals moved into the sea, gradually their legs became useless, and sometime in the Eocene—40 to 50 million years ago—there should have been whales with vestigial legs of declining usefulness for locomotion on land. One place where Eocene shallow sea strata are found is in Pakistan, and another is in the Egyptian desert southwest of Cairo. Bingo! The prediction was made, and in the past decade dozens of fossil skeletons of whales with legs have been recovered, in both localities.

A second recent dramatic prediction concerns the link between fish and amphibians, the much earlier step from sea to land. Existing fossils of fish and amphibians indicated that the transition should have taken place in the late Devonian period, that is, about 375 million years ago. Devonian rocks lie near the surface from Pennsylvania to

The excavation of the skeleton of a large *Basilosaurus* fossil whale with vestigial legs, at Wadi Al Hitan in the Western Desert of Egypt. The Eocene-age skeleton, including the skull, is about 16 meters, or 50 feet, long. A cast of the skeleton is on exhibit at the University of Michigan. Photograph by Philip Gingerich.

Kansas, and again in western United States and Canada, and they extend on north to the Arctic islands. However, not just any Devonian outcrops would be good hunting grounds, because only shallow seas near coastlines would be the likely habitat for transitional species. The more barren the area today, the greater the likelihood of actually finding relevant fossils, and Ellesmere island north of Greenland seemed like an appropriate target. During the sixth season of hunting in the desolate and chilly Arctic wasteland, a small team led by Neil Shubin of the University of Chicago struck pay dirt, not gold nuggets, but a transitional creature with fishy scales, a flat head with a neck, and fins bearing bones that would correspond to the limbs of a land-living animal. After consulting with the Inuit peoples' council of elders, the paleontologists choose one of their suggestions for a name: Tiktaalik, meaning "large freshwater fish." What was important about the find was that it came from both the predicted time frame and the predicted ancient environment.[15] A prominent "missing link" had been found.

In the past six decades a particularly conspicuous fossil gap has now been extensively populated. The biological family tree of *Homo sapiens* was essentially unknown in Darwin's day. Although the Neanderthal-type skull was found in Neander Valley in Germany two years before the *Origin of Species* was published, other early hominins (that is, any of the modern or extinct bipedal creatures) came only in the twentieth century. In 1924 Raymond Dart, an anatomy professor in South Africa, obtained a skull that to him seemed intermediate between ape and human skulls; he named it *Australopithicus africanus.* His claim was generally rejected until the late 1940s, by which time additional skulls had been found in South Africa. Beginning in the 1950s, other *Australopithicus* species were found in East Africa. The strata there included layers of volcanic lava, which contained radioactive potassium atoms. This circumstance has allowed the physical dating of those layers. In 1959 Louis and Mary Leakey found a hominin fossil in Oldavai Gorge in northern Tanzania, and by using the new potassium–argon technique, the fossil was dated to nearly two million years ago, a startlingly early period for paleoanthropologists of that time to consider. In the past half century numerous other hominin species have been found in Africa, including several in the genus *Homo.* I shall return to these presently, but first let us notice yet another pivotal discovery from the 1950s.

This breakthrough was the deciphering of the structure of DNA, by James Watson and Francis Crick, and the recognition that its chemical structure must play a fundamental role in the genetics of heredity. The numbers involved are simply mind-numbing. There are between 50 and 100 trillion cells in the human body. Most of these cells—red blood cells being the conspicuous exception—contain in their nuclei submicroscopic strands of DNA, which, if unraveled from a single cell, would stretch two or three meters. If all the DNA in all the cells in the human body could be stretched out and laid end-to-end, the thread could go to Jupiter and back multiple times. The arrangement of nucleic acids along these threads is like a text equivalent to the entire *Encyclopedia Britannica.* This text gives the instructions for building the proteins used by the cells. The instruction for making each protein is a gene in the Mendelian genetic sense.

By the end of the twentieth century this understanding led to the Human Genome Project, the mapping of the chemical sequences of the genes on the chromosomes in human cells. At first the process was tediously slow, but it was increasingly mechanized, and there was a brisk race to the final goal as the century ended in 2000. And the result brought a stunning surprise. Less than 2 percent of the DNA in human cells consists of genes, which all code for body-building proteins, and altogether there are only about 21,000 of them. Twenty-one thousand genes might seem like a large number, but the millimeter-long worm *C. elegans,* beloved of experimentalists, has 19,000 genes. How can humans be so much more complicated than a tiny worm when we have only 2,000 more genes? As an organism, we humans are immensely more complicated, so this result was at the time the most shocking outcome of the Human Genome Project. We are only now becoming able to understand its context.

Today biogeneticists are discovering that our cells can make far more than 21,000 proteins by splicing together pieces from existing genes to make numerous additional protein molecules. But where are the instructions for this? More than 98 percent of human DNA is *not* coding for genes. At first this vast additional array of nucleic acids was sometimes called junk DNA, a singularly inappropriate appellation because recent research is discovering that critical control functions, in the form of chemical switches, are embedded in this mass of nongenetic DNA. This understanding is today's field of epigenetics, literally "beyond the genetics of genes." This field will add a huge new vista to the mechanisms of evolution, because a mutation in the control system can cause a significant change in the structure of an organism. Natural selection is still at work, but the mutated material on which it works is undoubtedly more pliant than Darwin or the mathematical evolutionists of early twentieth century ever imagined.

As new species of hominins were found during the past six decades, and dated, there was an irresistible urge to connect the dots and place these species in a linked sequence between apes and the genus *Homo*. Today, largely through the efforts of Ian Tattersall at the American Museum of Natural History in New York, researchers recognize that several of these species are not ancestors of genus *Homo*, but belong on side branches of the hominin family.[16] Furthermore,

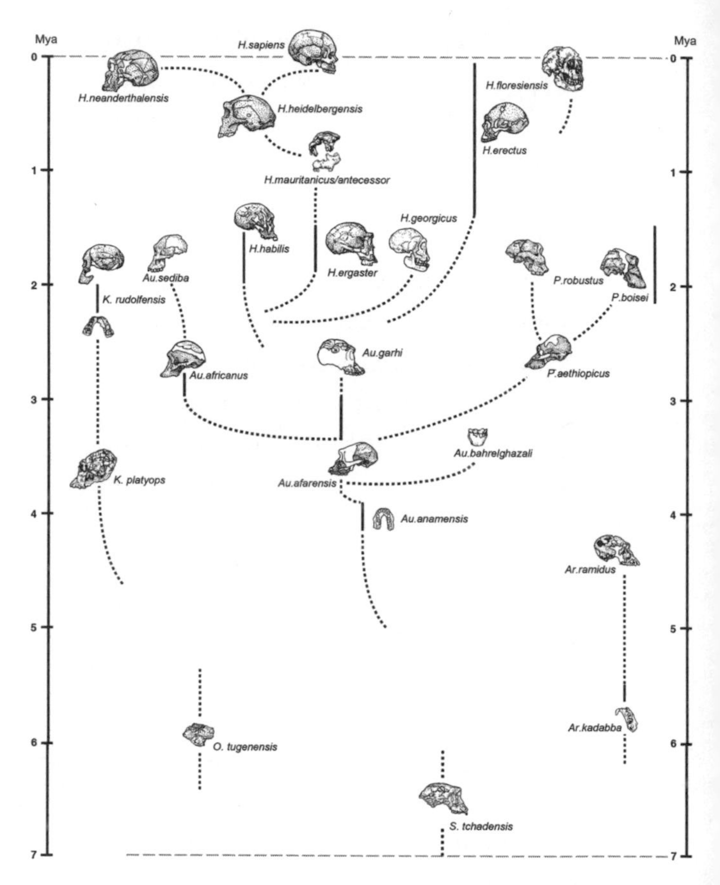

The family tree of *Homo sapiens* as of 2011. Ian Tattersall's chart shows the connections between the bipedal hominins. Courtesy of Ian Tattersall.

it was realized that during the past few millions of years several different species of *Australopithecus* and *Homo* were often likely to have been on Earth simultaneously.

All these life forms, whose existence was basically unsuspected a century ago, are scientific discoveries quite independent of evolution. They were, however, predicted by evolutionary theory, and the urge to place them on a family tree is a consequence of the theory of evolution. If we are looking for a scientific setting for the Genesis story of Adam and Eve, this is what we must cope with. Our modern knowledge of the human genome and of the chimpanzee genome indicates a close kinship between apes and humans, and a vast time scale. This is not evolution, but the scientific background for evolutionary theory.

The archaeological view is essentially a gradualist picture, with change coming slowly over grand vistas of time. In contrast, the Biblical picture is one of abrupt, rapid events. How long Adam was in the paradisiacal Garden of Eden is rather problematic. He was created one day and two days later he was naming the animals, a task that is still keeping today's taxonomists busy. Clearly the gradualist scientific picture is at odds with the swift origin of humankind depicted in the Genesis account. Can we reconcile these differing visions of creation by planting a sudden mutation event into the scientific picture? Or can we adapt a metaphorical view of Genesis into a long, gradual emergence of conscience and self-consciousness, two of the essential characteristics of humanness? The solution will involve our understanding of the role of Adam and Eve, with thoughtful contributions from both scientists and theologians.

I first began to hear rumors about the mitochondrial Eve in the 1980s, possibly even before the first pioneering paper was published in 1987. The mitochondria are organelles found in most human cells—red blood cells being the most conspicuous exception. They generate the chemical power for the cells and provide many cellular control functions. The mitochondrial DNA was discovered in the 1960s. Compared to the nuclear DNA in our chromosomes, which code for approximately 21,000 genes, the mitochondrial DNA codes for only 37 genes. But what is fascinating about the mitochondrial DNA is that it is transmitted only through the maternal line. What was discovered in the mid-1980s was that all mitochondrial DNA, worldwide, is essentially identical, as if it came from a single female ancestor of the entire human race, who was quickly christened the "mitochondrial Eve." I say "essentially identical," because in fact there are small variations, caused by mutations that have occurred over the ages, which made it possible to develop a family tree of sorts and to estimate when the mitochondrial Eve lived. This then led to a date for the mitochondrial Eve of about 200,000 years ago. Subsequently a similar analysis of the Y chromosome, which is transmitted only through the paternal line, led to a very similar date for the primeval Adam, a revised result announced in *Science* magazine in 2013.[17]

But were they alone? Or were they members of a much larger tribe of a few thousand creatures? A landmark paper published by geneticist Francisco Ayala in 1995, which deals with the genetic archaeology that has led over millions of years to *Homo sapiens,* concluded that the mitochondrial Eve could not have been the only female in the garden.[18] The pattern of subsequent mutations will not work out with such a severe bottleneck in the genomic tree.[19] The estimates of the size of the breeding population are highly variable, but generally in the vicinity of a few thousand individuals.

I will return to Adam and Eve shortly, but for now let me pick up the archaeological thread. The story of early human ancestors is not simply a story of skulls and bones. Moving from hundreds of thousands of years ago to tens of thousands of years, we find the fabulous prehistoric cave drawings in Lascaux and a few hundred other sites in southern France and Spain. The oldest cave art is in Chauvet (northwest of Marseille), discovered by speleologists in 1994; the Chauvet cave contains hundreds of spectacular animal drawings. Carbon-14 dating places the oldest drawings at 32,000 years ago. The drawings at Lascaux, with their striking images of bulls, deer, and horses, are approximately 20,000 years old.

Fifty thousand years is typically given as the awakening of modern *Homo sapiens sapiens.* Often the origin of language is given as early as 200,000 years, roughly the time of the mitochondrial Eve. But of course language doesn't fossilize, so it is difficult to pinpoint it. In any event, anthropologists assume that by 50,000 years ago language had reached a new level of sophistication.

At some time there must have been a cluster or series of mutations that made *Homo sapiens* physically different from his hominin cousins. But there seem to be no anatomical differences between early *Homo sapiens* and the modern *Homo sapiens sapiens* with his powers of language and art. As theologian Wentzel van Huyssteen points out, "There is an unbroken continuity between human and nonhuman brains, and yet at the same time there is a singular discontinuity between human and nonhuman minds, between brains that use this form of communication [that is, language] and brains that do not."[20]

Today the chance of locating some astonishing mutation that abruptly divides modern man from the earlier hominins seems remote. And it is, of course, the abrupt scenario that characterizes the Genesis story. Over the centuries a large theological edifice has been founded on both the nature of God and the nature of humankind as revealed in the Scriptures. Adam and Eve are seen theologically as God's special creation in the image of God, distinguished from the animals by the possession of souls, and of free choice, which promptly resulted in a wrong choice, bringing about the fall and original sin. Although science does not find genetic evidence for a single primeval pair from which we are all descended, epigenetics could at some point affect a local population out of the few thousand *Homo sapiens* existing at the time. The change could be subtle but profound and totally unrecorded in any possible fossil record.

When Pope John Paul II spoke to the Pontifical Academy of Sciences in 1996, he stated that "new findings lead us toward the recognition of evolution as more than a hypothesis," and without mentioning Adam explicitly he went on to say, "With man, we find ourselves facing a different ontological order—an ontological leap, we could say. . . . The moment of passage into the spiritual realm is not something that can be observed with research in the fields of physics and chemistry—although we can nevertheless discern, through experimental research, a series of very valuable signs of what is specifically human life. But the experience of metaphysical knowledge, of self-consciousness and self-awareness, of moral conscience, of liberty, or of aesthetic and religious experience—these must be analyzed through philosophical reflection, while theology seeks to clarify the ultimate meaning of the Creator's designs."[21]

In other words, John Paul was willing to take the evolutionary picture seriously, with the kinship between man and animals, and the gradual metamorphoses that separated them, but he also explicitly recognized the deeply significant transition to a spiritual being, a transition that does not fossilize. This experience of metaphysical knowledge must be probed through philosophical analysis, he said, to which we could add that it could well include mythmaking to convey theological insights. Such mythical interpretations, as opposed to historical ones, could be couched in a dramatic staging, where events of long ages could be condensed into a day or a week. If the early chapters of Genesis are not historical, it does not mean they are false or unimportant with regard to their theological insights. Truthful drama, but not actual history.

When I posed the question "Was Darwin right?" I knew there were unresolved theological issues, but closer to front and center is the issue of random chance versus design. When Darwin published his *Origin,* William Paley's famous book, *Natural Theology; or Evidences of the Existence and Attributes of the Deity,* was a foundational apologetic work, emphasizing the evidence of design. Darwin's book had none of this. A reader so inclined could conclude that Darwin's theory implied random variations as the efficient cause, with no designer's hand at work. Asa Gray perceived this as a problem, and in a review in the *Atlantic Monthly* stated, "There are only three views of efficient cause which may claim to be both philosophical and theistic."[22]

The first was that, at the beginning of time, matter was endowed with the properties to produce the phenomena. This was similar to the conclusion the essayist and Anglican minister Charles Kingsley reached soon after the publication of the *Origin* and which he later expressed in a sermon: "We knew of old that God was so wise that he could make all things; but behold, He is so much wiser even than that, that he could make all things make themselves."[23] Kingsley wrote to Darwin that "I have gradually learnt to see that it is just as noble a conception of Deity, to believe that he created primal forms capable of self development into all forms needful . . . , as to believe that He required a fresh act of intervention to supply the lacunas which he himself had made."[24] Darwin was very pleased with what Kingsley wrote, and so he revised the conclusion of his second edition to include those remarks, albeit anonymously. Darwin, always a sensitive stylist, edited the last line to read, "as to believe that He required a fresh act of creation to supply the voids caused by the action of His laws."

Quoting again from Gray's list of three views of efficient cause: "2. This same view, with the theory of insulated [isolated] interpositions, or occasional direct action, engrafted upon it—the view that events and operations in general go on in virtue simply of forces communicated at the first, but that now and then and only now and then, the Deity puts his hand directly to the work.

"3. The theory of the immediate, orderly, and constant, however infinitely diversified, action of the intelligent efficient Cause."

Let me quote further from Gray:

"It must be allowed that, while the third is preeminently the Christian view, all three are philosophically compatible with design in Nature. The second is probably the popular conception. . . . To silence his critics, this is the line for Mr. Darwin to take; for it at once and completely relieves his scientific theory from every theological objection which his reviewers have urged against it."

Concerning Gray's *Atlantic Monthly* article, Darwin replied that it was "admirable," but went on to respond, "I grieve to say that I cannot honestly go as far as you do about Design."[25] Subsequently he wrote, "I must think that it is illogical to suppose that the variations, which Nat[ural] Selection preserves for the good of any being, have been designed. But I know that I am in the same sort of muddle . . . as all the world seems to be in with respect to free will, yet with every supposed [action] to have been foreseen or preordained."[26]

Darwin was right to feel muddled by Gray's use of the word "design," something that philosopher Mortimer Adler has called the central error in Christian apologetics. Adler has argued that a predesigned universe would offer no place for freedom and choice. Twenty years ago Adler critiqued a piece I had written for *Great Ideas Today*, pronouncing mine an "excellent essay," but countering that I had skirted dangerously close to that central error.[27] I conceded that I should have used *purpose* instead of design. If Gray had emphasized intention rather than design, I think Darwin would have been more willing to accept Gray's arguments.

Darwin concluded his treatise with the lofty statement that "There is grandeur in this view of life, with its several powers, having been originally breathed into a few forms or into one; and that, whilst this planet has gone cycling on according to the fixed law of gravity, from so simple a beginning endless forms most beautiful and most wonderful have been, and are being, evolved."

The comparison with the law of gravitation was surely deliberate, for this was a law of nature accepted by all theologians, just as Darwin hoped his law would be accepted without invoking a theological discussion. In the second printing of his book two months later, to the dismay of atheists ever since, Darwin added the words "by the Creator" to make the last sentence read, "having been originally breathed by the Creator into a few forms or into one." The natural theologians of the day accepted Newton's law of gravitation, albeit set in motion by God, and no doubt Darwin wished to make more obvious the implied parallel between gravity and evolution.

Now in addressing question "Was Darwin Right?" I have attempted first to outline the foundation that Darwin himself had in formulating his *On the Origin of Species by Natural Selection.* Then also I have outlined how the idea of evolution of species has itself evolved as our studies of the age of the Earth, the fossil record, the nature of heredity, and the genetic kinship of life forms have all enriched our understanding of the development of life on Earth. In the first chapter I emphasized the key role of coherence of the resulting picture that is an essential ingredient in what constitutes a persuasive scientific understanding. Certainly the combined roles of geology, paleontology, chemistry, and biology have provided strong credentials for evolution today. I'm personally convinced that Darwin was on the right track. Yet a substantial fraction of Americans are reluctant to accept the evolutionary picture. What is missing?

Clearly the theory of evolution produces widespread tension. Central to this unease is the role of human identity in the evolutionary picture. Are we merely accidents of unguided random mutations, a glorious unexpected outcome of chance alignments of particles? What Darwin did *not* include in *On the Origin of Species* was where mankind fit within the evolutionary picture—that was too sensitive a topic. And that, I daresay, is still a sticking point today. But I believe that front and center of the resistance to evolution are the commentators who declare that evolution is tantamount to atheism.

It was within the milieu of evolution that Stephen Jay Gould framed his concept of nonoverlapping magisteria, the idea of two cultures, each minding its own business and going its own way by its own rules. And to a large extent this makes sense. Science and religion are two major cultural identities in the world today, and they use evidence far differently although their goal of a coherent understanding is remarkably similar. It is best if each can respect the other and cooperate rather than compete. But it is a fallacy that there is no overlap of magisteria. In particular, the fascinating question of how and when we became human inevitably offers an overlap and potentially competing world views.

One of the most brilliantly memorable summaries of evolution, by the eminent paleontologist George Gaylord Simpson, is a spectacular example of the overlap of magisteria. He wrote, "Man is the result of a purposeless and materialistic process that did not have him in mind. He was not planned."[28] Here is the magisterium of science, the mechanistic process of evolutionary change based on a long series of mutations that transformed life on Earth. It is, by its accepted rules, purposeless. To this Simpson has brought a personal interpretation out of an overlapping magisterium that deals with ultimate causes and purposes. But it will not do to take a working rule of the scientific magisterium as a personal ethos and pretend that is all there is. Simpson has simply superimposed a private interpretation onto the science, an interpretation that the science itself by no means demands.

For many people, the idea of science without God is an oxymoron, a self-contradiction. They may well look at Simpson's declaration and say to themselves, "If evolution requires me to be an atheist, then I'm having none of it!" I cannot help but wonder what fraction of the large number of Americans who do not accept evolution take their stance from statements of this sort.

Recently the engaging young evolutionary biologist from the University of Chicago, Jerry Coyne, returned to his alma mater, the Harvard Museum of Natural History, to give an articulate sixty-minute defense of the theory of evolution. He then added a ten-minute peroration on why evolution is so widely denied in America. He showed a bar graph in which only Turkey demonstrated more disbelief in evolution than America.[29] At the top of the graph were those who were the most accepting of evolution: Iceland, Denmark, and Sweden. These Scandinavian countries were rapidly giving

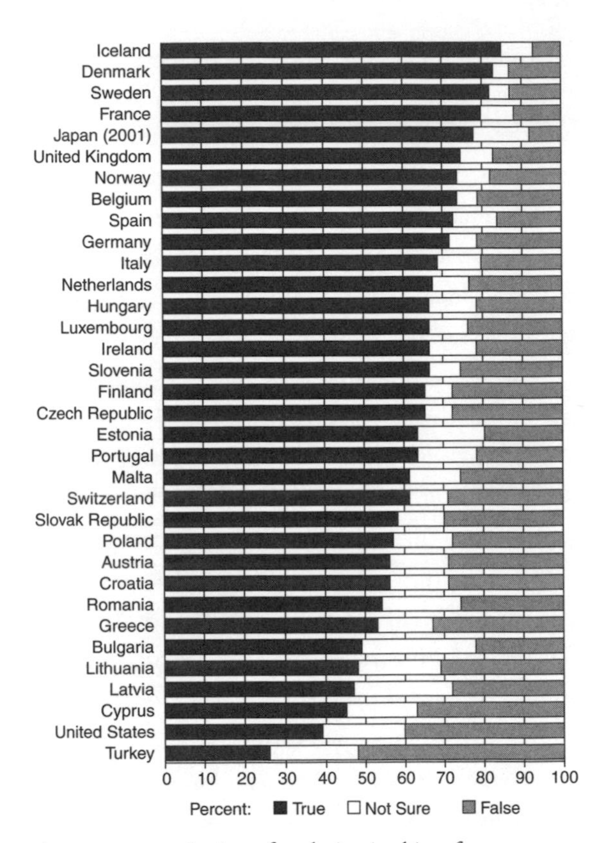

Acceptance or rejection of evolution in thirty-four countries, 2005.

up religion, he declared, and we should try harder to emulate them. "Evolution demands atheism" was his screed.

I shook my head in astonishment. How could the man believe that his attack was winning converts for evolution? Here was a spectacular case of an overlapping magisterium of personal disbelief wagging the dog. His lecture and his blog are not ways to get a sympathetic hearing for one of the most impressive edifices of modern science.

I am sympathetic if Professor Simpson, based on his insights into evolution, was led to his conclusion that we are the unplanned results of myriad mutations. I just think his was is an incomplete view. I am more inclined to accept Freeman Dyson's opinion that it looks as if this is a universe that knew we were coming.[30] I think there are a few more things we must consider for a coherent view of the cosmos, which I will reflect on when I ask, "Was Fred Hoyle right?"

Fred Hoyle (1915–2001). Photograph courtesy of the Institute of Astronomy, Cambridge, England, where Hoyle was founding director.

3

Was Hoyle Right?

I NEVER MET Copernicus, but I did have the unusual privilege of attending his funeral. This occurred in 2011 when his bones were reinterred with suitable pageantry at the Frombork Cathedral in northern Poland.[1]

Likewise I never met Charles Darwin, though I feel some indirect kinship with him every day when, on the way to my Observatory office, I pass the house of Asa Gray, his leading American correspondent and supporter.

But Fred Hoyle, the quintessential outsider and the pivot of this third chapter, I met repeatedly. Nevertheless, I was young and junior then, and today I can think of tough questions I never asked him.

In this sequence of reflections on the nature of science, I have argued that what passes as acceptable science depends on the coherence of its vision of the universe. This picture is drawn as a very large tapestry, a vision that includes resonance with a broader cultural milieu. Science today is one of the largest cultural components, religion another, and sometimes their relations are deeply fraught. My late colleague Stephen Jay Gould proposed what he called nonoverlapping magisteria as a way to solve potential conflicts between science and religion. They can be friends, he argued, if they stayed out of each other's domains. In some respects that is a good idea, but the reality is that the magisteria overlap in powerful ways that may not always seem obvious to the participants.

In this chapter I propose to address two contemporary arenas where the players themselves seem unaware of the outside assumptions they are bringing into their science. I call it metaphysics, which literally means "beyond physics," because unwittingly their assumptions come from a magisterium lying beyond physics. (And yes, I know that originally the term meant "next to *Physics*" in the library of Aristotle's works, but the term has gradually evolved over the millennia.)

I have chosen Fred Hoyle, one of the greatest astrophysicists of the last century, as a modern archetype because he wrestled with some of these issues even though he was only peripherally involved in the two areas I will focus on in the latter part of this chapter: multiverses, and, briefly, intelligent life on other worlds.

Fred Hoyle was born in 1915 in the countryside of west Yorkshire eloquently described in the novel *Wuthering Heights.* In 1933 he entered Emmanuel College Cambridge, coming from an impoverished family background and with a distinctly non-university Yorkshire accent, and thirty-nine years later left Cambridge in a misguided huff. But in between he ascended into the highest ranks of British science, becoming vice-president of the Royal Society and president of the Royal Astronomical Society, almost single-handedly returning Britain to the top echelons of international theoretical astrophysics, and setting it on the path toward excellence in observational astronomy. It is a stirring story worthy of Charles Dickens, of an inquisitive, rough-hewn lad making the grade in the tightly traditional world of Cantabrigian academia, yet with the depths of a Greek tragedy in which the flawed hero finally becomes an outcast.[2]

Fred Hoyle first came into my purview in 1950 when, as a young editorial assistant at *Sky and Telescope* magazine, I read about Hoyle's *The Nature of the Universe* in the scathing review written by Frank Edmondson of Indiana University. I learned there that Hoyle was a proponent of a dubious steady state cosmological theory in which the universe had lasted forever with no beginning, but I did not then discover that he had just invented a possibly derisive term, "big bang," to characterize the leading competitive cosmology. The last time I saw Fred was at the American Philosophical Society in Philadelphia in November 1981, when he had come to sign the roll book of this venerable American academy. Before that, I had occasionally seen him in the English Cambridge or at international meetings.

The genesis of Hoyle's popular book, that which first made him famous, was a series of seven radio talks given on the British Broadcasting Corporation in 1950. Today that book, *The Nature of the Universe,* seems oddly dated. Nevertheless, it is still of note for its two most controversial chapters: one on the steady state cosmology and the final one, which includes his hostile remarks on traditional Christianity, seen as gratuitous by his critics. Let me first give some historical context to the state of large-scale cosmology in 1950.

Curiously, modern cosmology was stillborn with the work of Isaac Newton. Imagine a universe with stars sprinkled through space, each a gravitational attractor. Unless the stars are carefully placed so that the gravitational attractions on to each star are balanced out, the whole thing will fall in on itself. The entire universe would be intrinsically unstable. Displace one star slightly and there will be more gravitational attraction on one side than the other, and the imbalance will generate an increasingly rapid runaway collapse. The mathematician David Gregory reported hearing Newton say that a continual miracle was needed to prevent the Sun and fixed stars from rushing together as a result of gravity.[3] It is no wonder that cosmology was dead at its birth.[4]

Now fast forward to the 1920s, with Einstein's general theory of relativity, a new theory of gravity.

And here the same problem occurred. To prevent his theoretical universe from collapsing in on itself, Einstein was obliged to introduce a repulsive term into the mathematics, a cosmological constant to keep his mathematical universe stable. However, unknown to Einstein, the universe was not a vast distribution of stars, but of distant galaxies, a fact established by Edwin Hubble in 1924. And then, by 1928, the work of Abbé Georges Lemaître and Hubble showed that the galaxies were rushing away from us in an expanding universe. The farther they were, the faster they were going. It seemed the cosmological constant was no longer needed to keep it quiescent, because the universe was explosively expanding. Einstein reputedly said that the cosmological constant was his greatest blunder, but he never wrote that statement down and it may just be a mythic quotation.[5] Ironically, the recent Nobel-winning work of the late 1990s has brought the cosmological constant back into play to account for the acceleration of the expansion.

One consequence of the expansion of the universe is that we can calculate an age, the time of the initial explosion that started it all off. A simple

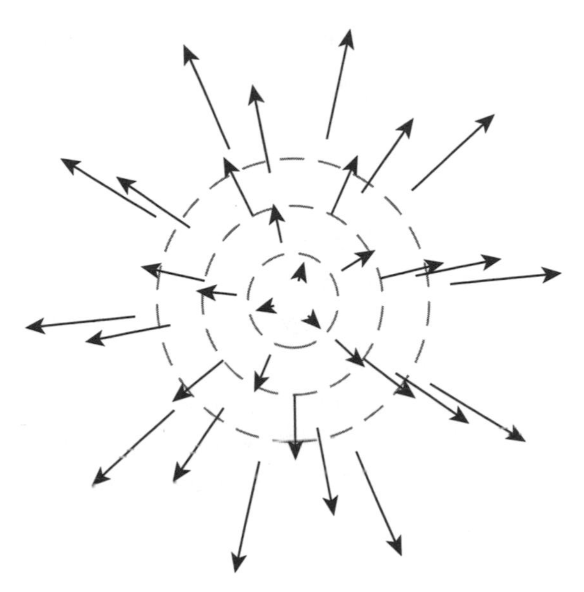

Schematic diagram of galaxies expanding into newly formed space following the Big Bang.

analogy will show how this works. Suppose that I was giving a lecture at Gordon College, about thirty miles north of Boston. Suppose also that when the lecture ended, each audience member headed out in a straight line and at a constant speed, scattering in every direction from the hall. At some later time we collect the data on each person's distance from the hall and the speed of travel. Some, filled with questions, may still be lingering in the hall. Others, in speeding sports cars, may be all the way to Boston. An attendee with a private helicopter might be nearing Dartmouth in New Hampshire. Given their speeds and distances, it would be a simple calculation to know when the lecture ended.

Likewise, back in 1950, astronomers could estimate the distances to the galaxies—the fainter they were, the farther they were. And the speeds could be determined from the Doppler red shifts of the lines in their spectra. The distance–speed relation showed that the universe was approximately a billion years old. Now a billion is a pretty big number. Today we are getting used to numbers of this size. The population of the world is approaching 8 billion persons. The annual Pentagon budget for nuclear weapons is some tens of billions of dollars. To count to a billion, a number every second, day and night, would require thirty-one years. But a billion years did not seem like a long enough time frame to account for the ages of the oldest rocks, or for some of the astrophysical processes to take place. This was very awkward, having a universe younger than its parts.

One way out of this dilemma was to consider a cosmology with no beginning. In his book *The Nature of the Universe,* Fred Hoyle contrasted this idea with "the older ideas" from the 1920s and 1930s. "One of them is distinguished by the assumption that the universe started its life a finite time ago in a single huge explosion. . . . This big bang idea seemed to me to be unsatisfactory even before detailed examination showed that it leads to serious difficulties."[6] That statement from his BBC radio series was the very first use of the expression "Big Bang." Hoyle did not say that a principal reason the Big Bang model was unsatisfactory was that it was too much like the Genesis account of God saying "Let there be light!," although I suspect that was not far from his mind and those of his colleagues who came up with the idea of continual creation.

Now the starting simplification of any mathematical cosmology was to assume a homogeneous distribution of matter, that is, all the stars and galaxies smoothed out into some uniform density.

This working assumption was called the cosmological principle. Hoyle's colleagues, Hermann Bondi and Tommy Gold, two refugees from Europe who had worked with Hoyle during the Second World War, proposed there should be what they called the perfect cosmological principle, namely that the universe should be homogeneous and uniform not only in space, but also in time. But how could that be? If the universe was expanding, the matter should be getting thinner and the density should drop, so the universe would not have a constant density in time. But suppose that somehow hydrogen atoms were being spontaneously created, so to say, in the cracks of the universe, adding just enough mass to keep the density constant. As Hoyle pointed out, the average rate of appearance of matter would amount to the creation of one atom per year in a volume equal to St. Paul's cathedral. Obviously unmeasurable.

Hoyle's radio talks not only highlighted a variety of his own speculations that were far from being widely accepted—including of course the so-called steady state cosmology maintained by continual creation—they presented a final "personal view" that threw down the gauntlet to fundamental Christian beliefs. "Here we are in this wholly fantastic Universe," he remarked, "with scarcely a clue as to whether our existence has any real significance."[7] He went on to say, "It strikes me as very curious that the Christians have so little to say about how they propose eternity should be spent. . . . Now what the Christians offer me is an eternity of frustration." He added that he thought 300 years would be enough.

The British Broadcasting Corporation asked several persons to respond to Hoyle's provocative remarks, including Dorothy Sayers who was probably better known as the author of the Lord Peter Wimsey mysteries than as a Christian apologist. But she was very active in the circle of C. S. Lewis, whose own BBC talks had generated the best selling book *Mere Christianity*. In remarks under the heading "The Theologian and the Scientist," Sayers noted that "when modern scientists begin to discuss religion, I often wish that some kindly soul had thought of sending them to Sunday School. For they do not seem to know the meaning of the words that Christians use. Here, for example is

Mr. Fred Hoyle. He finds the idea of immortality 'horrible' because he himself would not care to live more than 300 years. And he complains that Christians 'have so little to say about how they propose that eternity shall be spent.'

"Christians have, in fact, said a good deal about the nature of eternal life—in particular that it does *not* consist (as Mr. Hoyle seems to think) of endlessly prolonged time of the kind we know. They insist that, although we are often obliged to picture eternity in terms of time, the two things are really incommensurable."[8] Sayers went on to use the analogy of a novelist, over an undisclosed length of time developing a character, whose entire trajectory is then on display simultaneously, the difference between eternity and immortality.

In the next few years that followed Hoyle's BBC presentation of continuous creation, a major revolution occurred. Walter Baade at Mount Wilson and Palomar Observatories discovered that there was not just one but two kinds of Cepheid variable stars, which had provided the distance scale to the Andromeda galaxy. When that was sorted out, astronomers recognized that the whole distance scale, and hence the cosmological age calculations, were out by a factor of two, so the problem of the universe being younger than its parts went away, at least for the moment.

In 1951 Pope Pius XII declared that the Big Bang cosmology gave evidence for creation and a Creator, though he didn't expressly use the term Big Bang as that had not yet gained currency as the universal, popular name for the anti–steady state cosmology. His triumphalist address to the Pontifical Academy of Sciences concluded, "Thus, with the concreteness which is characteristic of physical proofs, [science] has affirmed the contingency of the universe and also the well-founded deduction as to the epoch when the world came forth from the hands of the Creator. Hence Creation took place. We say, therefore, there is a Creator. Therefore, God exists."[9]

The waggish George Gamow, a full supporter of the Big Bang theory because of his theoretical conception of the hot origin of the elements, in a characteristically whimsical mood inserted a footnote in one of his technical papers citing the papal statement as infallible proof that his cosmology was now beyond all doubt.[10] In contrast, Georges Lemaître, then still active as a Catholic cosmologist, was dismayed by the Pope's entangling of physical cosmology and theological creation, and made sure that the director of the Vatican Observatory had a few words with Pope Pius. As astrophysicist and Anglican priest Rodney Holder has put it, both Fred Hoyle and Pius XII had confused theological Creation with astrophysical origination.[11] In any event, from the atheists' point of view, the papal statement was like pouring gasoline on the smoldering controversy.

The 1950s saw the development of radio astronomy, with the discovery of celestial sources emitting radio noise, originally called "radio stars" because they were unresolved and twinkled. After some dispute, most of them turned out to be extragalactic, that is, lying beyond our Milky Way galaxy; by and by astronomers realized in many instances that the radio noise emanated from distant sources, eventually called quasars. (The quasars turned out to be distant and therefore ancient galaxies with extremely active and luminous nuclei powered by black holes.) There arose a fearsome ideological controversy between Fred Hoyle, who was defending the steady state cosmology, and his fellow Cambridge scientist, radio astronomer Martin Ryle, about the significance of the data that were pouring in. Indeed, there were genuine problems with some of the earlier data, but it eventually became clear that the observed universe was *not* homogeneous in time—rather, it was a universe with a history. The more distant the observations, the earlier the time frame, and looking outward in both space and time there was an epoch when quasars flourished and were far more abundant than they are now. Out beyond that, going further back in time, there was a sparsity of quasars. And then came the crowning blow, the discovery in 1965 of the so-called cosmic microwave background radiation, the evidence of the primordial fireball from which our universe sprang.

It was time to give up. Fred Hoyle was wrong. And in September 1965, in a lecture to the British Association for the Advancement of Science, he announced that he had erred, lovely as the continuous creation scenario had seemed to him. Steady state cosmology was dead.

But the steady state cosmology had posed an interesting problem that did not go away when that cosmology died, and it is Fred Hoyle's key contribution to its solution that keeps his name high in the annals of twentieth-century astrophysics. If hydrogen were continually being created in the cracks of space, where did the heavier elements come from? In the alternative Big Bang cosmology, were the more massive elements produced in the intense nuclear cataclysm of the initial few minutes? The physicist George Gamow had long proposed such a scenario, but detailed calculations failed because of the lack of a stable nucleus of mass 5, seemingly required as a stepping stone for building heavier elements, such as carbon, with a mass of 12, or oxygen, with a mass of 16. Thus from the Big Bang only hydrogen (mass 1), helium (mass 4), and a tiny percentage of lithium were produced, so that the path to the heavier elements remained a mystery. In particular, carbon was critical for the formation of life.

Carbon is essential for the complexity of life because it can form so many different and often intricate molecules. Unlike almost every other element, it can easily bond with itself, forming chains or rings.

Consider the number of molecules that can be made just by joining hydrogen with an element. For example, with oxygen, we get just two molecules that combine oxygen with hydrogen: H_2O (water) and H_2O_2 (hydrogen peroxide). But for self-bonding carbon, the tally of different molecules made just out of carbon and hydrogen is overwhelming.

Li	Be	B	C	N	O	Fl	Ne
1	1	7	~2300	7	2	1	0

Because this table is taken from my chemistry notes of six decades ago, the numbers for carbon need updating. A Google search gives "thousands," "vast," and "near infinite" as the number of hydrocarbons. Whatever the current number, it is clear that carbon greatly exceeds any other atom (except hydrogen) in the number of different molecules it can make. Because of its self-bonding properties, only carbon compounds can achieve the complexity required for living organisms.

Darwin wrote about the survival of the fittest. Exactly a century ago the Harvard chemist Lawrence J. Henderson turned that around with a book entitled *The Fitness of the Environment.* Not only must life itself be robust enough to survive, but a favorable environment including the available chemistry was required. Henderson argued that without the high availability of carbon—and it was then not yet known that the carbon is the fourth-most abundant element in the cosmos—there would not be life on Earth. Similarly, lacking the unusual and remarkable physical properties of water, we would not be here. And Henderson, arguing from his own specialty (carbon dioxide and its solubility in water), noted that macroscopic animals such as ourselves would not be able to get rid of the waste from oxidation processes in our bodies—the source of our energy—without some unique properties of the carbon dioxide molecule. We breathe out about two pounds of carbon dioxide every twenty-four hours. The extraordinary solubility of carbon dioxide in water allows us to exhaust this waste product, but what is waste to us provides the building blocks for plants. Today we know that beyond these special properties cited by Henderson is a series of precisely tuned physical constants that make our environment and its chemistry fit for life. It is as if our universe has been designed for life.

However, in 1951, neither the Big Bang nor the steady state cosmology could explain how carbon could be formed from the abundant hydrogen that both theories postulated. Then, in 1952, Edwin Salpeter showed an alternative route that could take place over long periods in the hellishly hot interiors of evolving giant stars. But when Hoyle examined the process theoretically, he realized that, without some special help, it was too slow to produce the amount of carbon required for life on Earth.

Only if the carbon atom had what is called a resonance level at exactly the right place to enhance the production of carbon would Salpeter's route yield a suitable abundance of carbon atoms. Hoyle was at that time in Pasadena, so he suggested to Willy Fowler, a nuclear physicist, that he should look for the resonance level with the new Caltech Van de Graaff accelerator. Fowler thought it was a little crazy that this British theoretician could actually predict where a carbon resonance level would lie. As Simon Mitton put it in his biography of Hoyle, "Hoyle's claim was outrageous. The cosmologist was claiming he could do what no nuclear physicist in the world could

The outer casing of one of Willy Fowler's Van de Graaff accelerators at Caltech in the 1950s. The picture shows the scale of the instrumentation used to measure resonance levels in the carbon atom. Caltech Archives.

do: predict a precise energy state in an atomic nucleus."[12]

Mitton's account continues, "Fowler began to sense that, if Hoyle were right (highly improbable in his view), the consequences for element synthesis in stars would be immense. He rounded up his team for a council of war in his crowded office. Hoyle, surrounded by the world's brightest experimental nuclear physicists, suddenly became acutely aware that he could end up looking ridiculous. Fortunately for him, the nuclear physicists reached a consensus on how to run the experiment. First, they had to make a major experimental modification. They decided to run with a small particle accelerator, using a spectrometer then attached to their large accelerator. The spectrometer, a device for measuring the energy of particles emitted in nuclear reactions, would have to be moved. It weighed many tons on account of the huge magnet it contained, and had to be maneuvered down a narrow hallway four feet wide, and round two corners. They rested a large steel plate on several hundred tennis balls, slid the multi-ton instrument onto this platform, and set the whole in motion. A pack of graduate students feverishly fed the squashed tennis balls from back to front as the procession proceeded!"

The measurement was not simple and required running the accelerator for about three months. To Fowler's surprise the resonance level fell exactly where Hoyle had calculated it should be. Without this tightly placed resonance level in carbon, and a comparable one in oxygen, carbon would be too rare to play the essential role in the biochemistry of life.

Within a few years Hoyle and Fowler, together with Geoffrey and Margaret Burbidge, worked out in considerable detail how the building of the heavier elements could take place in the cores of evolving stars, and in 1957 they published one of the most important astrophysics papers of the twentieth century, which goes under their initials in alphabetical order, B^2FH. This was a massive paper requiring months of work. In a letter published in *Science,* Geoff Burbidge remarked that Hoyle alone formed the theory of stellar nucleosynthesis.[13] It is an important example of where Hoyle was right.

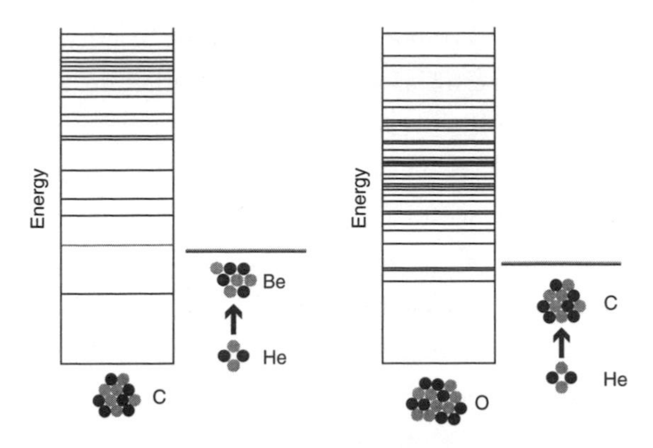

The arrangement of protons and neutrons in an atomic nucleus have different configurations depending on the amount of energy available. Only certain energy amounts, called resonance levels, lead to stable arrangements. The diagrams show these levels, discovered with nuclear accelerators, for carbon (C) (left) and for oxygen (O) nuclei (right). The protons (black) and neutrons (gray) are shown schematically for the various nuclei. When helium (He) bombards a beryllium (Be) nucleus, a small part of the mass is annihilated and converted into energy according to Einstein's $E=mc^2$ equation. The kinetic energy of the incoming particle raises the total energy just enough to form an excited but stable carbon nucleus. Subsequently (right-hand side of the diagram), a helium nucleus could bombard the carbon nucleus, but now the combined energies are too great compared to the closest resonance level in the oxygen nucleus. Enough oxygen is produced to make it more abundant than carbon, but the process is sufficiently inefficient to allow a substantial survival of carbon.

Eventually rumors flew that nothing had shaken Hoyle's atheism as much as his prediction and subsequent discovery of the resonance state of the carbon nucleus, which makes possible the substantial abundance of carbon and consequently the possibility of carbon-based life in the universe. From time to time I sat with Fred, generally discussing one historical issue or another, but I never had quite enough nerve to ask if this discovery had shaken his atheism. But here is what he wrote in the November 1981 issue of the Caltech alumni magazine: "Would you not say to yourself, 'Some supercalculating intellect must have designed the properties of the carbon atom, otherwise the chance of my finding such an atom through the blind forces of nature would be utterly minuscule.' Of course you would. . . . A common sense interpretation of the facts suggests that a superintellect has monkeyed with physics, as well as with chemistry and biology, and that there are no blind forces worth speaking about in nature. The numbers one calculates from the facts seem to me so overwhelming as to put this conclusion almost beyond question."[14] Not only was the resonance level in the carbon critically positioned to speed up the formation of carbon, but another such level in oxygen was just right to slow down the production of oxygen, which prevented much of the carbon from being transmuted into oxygen. As astrophysicist Paul Davies has put it, the little bear's porridge was just right.

The precise positioning of these two resonance levels are examples of what is today referred to as fine-tuning. In the final decades of the twentieth century cosmologists became increasingly aware of, and puzzled by, the fact that a number of physical constants of nature seem singularly tuned to allow the existence of intelligent life on Earth. Change the so-called fine structure constant by only a few percentage points and carbon (a virtually indispensable element for complex life) would become rare instead of number four in the list of most abundant atoms. Even small alterations in the numerical ratio of the kinetic energy in the Big Bang explosion to the gravitational potential energy of the mass then created, a result sometimes referred to as the "flatness" of the universe, would disrupt an array of physical processes with the result that stars and galaxies would not form. The astronomer royal, Martin Rees, has called this "the most remarkable feature of our universe."

Consider as well the huge ratio between electrostatic and gravitational forces, 10^{36}. If gravity were a million times stronger and the ratio were 10^{30}, a typical star would last only 10,000 years, and if evolution could go that fast, the strong gravity would limit the largest creature to something like the size of an insect.

A decade after his famous statement on the absence of blind forces in nature, Fred Hoyle expressed much the same view in closing a dinner discussion arranged by Ruth Nanda Anshen on the origin of the universe and the origin of religion: "The issue of whether the universe is purposive is an ultimate question that is at the back of everybody's mind. . . . And Dr. Anshen has now just raised exactly the same question as to whether the universe is a product of thought. And I have to say that that is also my personal opinion, but I can't back it up by too much of a precise argument. There are very many aspects of the universe where you either have to say there have been monstrous coincidences, which there might have been, or, alternatively, there is a purposive scenario to which the universe conforms."[15] Such talk, and the whole idea of fine-tuning, has been disturbing to the atheists. How can the magisterium of science cope with the idea that there is purpose in the universe? Not everyone is as candid as Fred Hoyle was.

We might call this the atheist dilemma. Either there are monstrous coincidences, or there is purpose in the universe. The universe-with-a-purpose scenario is very troubling to those who accept only a purely mechanical cosmos. One creative solution of this puzzle is to inquire whether there is only a single way to make a universe. We simply do not know if this is the case, but just suppose that there are multiple possibilities. If there are other possibilities, would not these alternative universes actually exist somewhere? That, by the way, is called the principle of plenitude, which has a long historical tradition, going back to Plato. In any event, this is the basis of the popular multiverse idea. And if there could be another universe, why not a few more universes, with assorted dimensions, physical constants, chemistry, and maybe even biology? Suppose there are ten universes. If in this cosmic roulette perhaps one would occur with the right parameters for life, then necessarily that is the one in which we would be found. All the fine-tuning would have to be just right or we would not be here to think about it. Still, there is a rather slim chance of getting everything accidentally fine-tuned just right with only ten possibilities.

Somehow this reminds me of the legend of the sorcerer's apprentice, who knows just enough of his master's magic to animate a broom to carry a bucket of water to fill the bath, but not enough to stop the process. Goethe's poem was converted into a lively musical score by Paul Dukas, and thence into a sequence in Walt Disney's *Fantasia* in which Mickey Mouse plays the role of the inept apprentice with the enchanted broom. Desperate to halt the oncoming flood, Mickey takes an axe to split the broom, which simply doubles the work force. Mickey strikes again and again, thereby producing a veritable army of water-carrying brooms, thoroughly dousing the hapless apprentice.

So why stop with just a few extra universes, especially when we discover hints that our life-bearing universe is a real outlier? Its properties, so congenial for the formation of life, seem to be distinctly unusual. But if a life-friendly universe is rare, why not imagine lots of possibilities? To some cosmologists the discovery of inflation, the incredible enlargement of the universe in its first split second, suggested that an inflating universe could spawn other universes just as Disney's sorcerer's apprentice could unwittingly create a plethora of enchanted brooms. It is possible to imagine the birth of multiple universes, but how to stop the process? Once it got started, there could be millions or trillions of universes, or even many, many more forming continuously.

We would never know just how unusual or miraculous our own life-friendly universe is. We would, like Shakespeare's Miranda, simply be in a position to exclaim, "O, brave new world that has such people in't." Or possibly, joining the Psalmist, "When I consider the moon and the stars, What is man that thou art mindful of him?"

As originally proposed, these universes were in their own separate spaces, totally disconnected from each other. And thus these additional universes would remain forever unobservable from our own universe. This arrangement seems so far beyond physics that I call the whole idea metaphysics. There is, however, another form of the multiverse theory, perhaps invented when the multiverse enthusiasts began to recognize the difficulties of considering as scientific the first, unobservable form of multiple universes. In this new form, which has risen like a phoenix from the ashes of the original theory, the various "universes" (with their entirely different physical laws) lie within the same space, but as regions with differing physical laws. Don't ask me what happens at the boundaries, particularly if one region is three-dimensional and another is six-dimensional, maybe with two dimensions of time! Enthusiasts speculate that different regions can collide, and that we might be able to discern visible consequences of collisions between regional environments with different tuning and physical laws. It is wild and wonderful speculation, very comforting to those who like to think that this finely tuned world is merely random chance.

The great flourishing of modern science has been built on *efficient* causes, how things work, deliberately suppressing the *final* causes, the why of how things work as they do, so favored by Aristotle. A simple example, popularized by John Polkinghorne (a professor of physics at Cambridge University who left to become an Anglican priest), distinguishes between efficient and final causes. He asks, "Why is the water in the tea kettle boiling?" and we can answer, the water is boiling because the heat from the fire raises the temperature of the water until the molecules move faster and faster so that some escape from the surface and become a gas. But we can also answer that the water in the tea kettle is boiling because we want some tea. The first answer illustrates what Aristotle called an efficient cause, an explanation of how the phenomenon takes place, whereas the second answer, "Because we want some tea," is a final cause, the reason the phenomenon takes place. One aspect of the scientific revolution of the seventeenth century was that it turned away from the final causes so central in the Aristotelian world view and concentrated on efficient causes, the how of the phenomena.

It was Sir Francis Bacon, in his book *The Advancement of Learning,* who railed against final causes.

Natural philosophy inquires into causes, he wrote: the one part, which is physics, inquires and handles the material and efficient causes, and the other, which is metaphysics, handles the formal and final causes.[16] Human understanding, however, while aiming at further progress, falls back to what is actually less advanced, he claimed, namely, final causes; for they are clearly more allied to man's own nature, than is the system of the universe, and from this source have wonderfully corrupted philosophy.[17] But surely, Sir Francis, if we wish to go beyond physics, to consider metaphysics, the reasons why, we need to think about final causes.

To me, belief in a final cause, a Creator God, gives a coherent understanding of why the universe seems so congenially designed for the existence of intelligent, self-reflective life. Only small changes in numerous physical constants would render the universe uninhabitable. I do not claim the fine-tuning as a proof for the existence of a Creator, only that to me, the universe makes more sense with this understanding, and that is the core of my belief.

Harumph! I can imagine a feisty atheist challenger declaring. You are falling into an age-old "God of the gaps" trap. When you cannot understand the physical or biological basis of some phenomenon such as the apparent fine-tuning, you simply say "God did it!," pack up your critical senses, and go home. You don't say how it was done!

What the atheist challenger is missing here is that I was arguing that the fine-tuning can be understood as the design of a Creator God as a formal cause, not the how of an efficient cause. I was working from the magisterium of religious and theological understanding, but this is clearly a case of overlapping magisteria because part of my theological understanding is that the heavens declare the glory of God, and that is reflected in my scientific understanding of fine-tuning. It is a matter of belief, not of proof. But the multiverse proponent is likewise arguing from belief, a final cause, in the absence of a satisfactory efficient cause.

Preeminent among the skeptics who oppose the multiverse idea is cosmologist George Ellis, professor at the University of Cape Town, and sometime collaborator with Stephen Hawking. He quotes *Scientific American* columnist Martin Gardner: "There is not the slightest shred of reliable evidence that there is any universe other than the one we are in. No multiverse theory has so far provided a prediction that can be tested. As far as we can tell, universes are not plentiful."[18] Earlier in his essay Ellis states, "The claim that multiverses exist is a belief rather than an established scientific fact. . . . Despite this, many articles and books dogmatically proclaim that the multiverse is an established scientific fact."[19]

There is that word "belief" again, but this time belonging to supporters of the multiverse theory. Many are atheists, but by no means all. It seems that metaphysical ideas from an overlapping magisterium have entered the scene again. In a modest rebuttal to Ellis's essay, University of London astronomer Bernard Carr writes that "without a multiverse one may be forced to adopt a nonphysical explanation like a fine-tuner, which is why Neil Manson [a philosopher at the University of Mississippi] claims that 'the multiverse is the last resort of the desperate atheist.'"[20]

The imagined and imaginary multiple universes have a curious parallel in the history of science of over a century ago, as was pointed out by historian of science Ted Davis in commenting on the lecture that was the basis of this chapter. James Clerk Maxwell, perhaps the greatest physicist of the nineteenth century, derived from the known properties of electricity and of magnetism the equations for electromagnetic waves, including light itself. His equations predicted the possibility of radio waves, which were discovered by Heinrich Hertz in 1888, fifteen years after Maxwell had published his theory. Because the existence of electromagnetic waves seemed to require some medium to be undulating, Maxwell prepared perhaps the most remarkable encyclopedia article of the century, on the luminiferous ether, for the ninth edition of the *Encyclopaedia Britannica.* The ether was, he wrote, "a material substance of a more subtle kind than visible bodies, supposed to exist in those parts of space which are apparently empty." As Davis pointed out, physicists in Maxwell's day still believed in a version of something originally proposed by Aristotle—the stuff heaven is made that heaven is made from a fifth element, quintessentially pure and perfectly changeless, unlike any of the four elements that make up everything below the sphere of the moon. Why did Maxwell still believe in something like this, a unique type of matter filling all space in the universe? Why did he not think that empty space is just—well—empty?[21]

The answer is quite simple: light is a type of wave motion, and waves require some sort of physical medium through which to travel. They cannot travel through truly empty space—or so Maxwell's contemporaries assumed. But, as Lord Salisbury (an adept in physics who was just about to become prime minister of Britain) wrote in 1894, if light is an undulation, "there must be something to undulate. In order to furnish that something, the notion of the ether was conceived, and for more than two generations the main, if not the only, function of the word ether has been to furnish a nominative case to the verb 'to undulate.'"[22] Thus, there was the virtually unanimous assumption that ether must exist.

Maxwell had reinforced the belief in ether with his encyclopedia article, where he declared, "The evidence for the existence of the luminiferous ether has accumulated as additional phenomena of light and other radiations have been discovered; and the properties of this medium, as deduced from the phenomena of light, have been found to be precisely those required to explain electromagnetic phenomena."[23] Nevertheless, no one succeeded in detecting the ether with a measurable result, which is why ether has disappeared from scientific journals. Time will tell if the multiple universes will meet the same fate.

But one can simply withhold judgment with respect to the multiverse theory, as Steven Weinberg, who is an articulate and very thoughtful atheist, has done. A decade ago, I moderated a debate between him and John Polkinghorne on whether the universe was designed. I asked Steve on what basis beyond faith can one justify the idea of multiple universes. "I don't maintain that that idea is true," he responded. "It's a possibility that has emerged and it remains a possibility. When I become convinced of its truth, it will be because the equations of physics that unify the various forces—the equations of quantum mechanics, relativity, all that—have that as a consequence. It won't be an act of faith. It will be a deduction from laws which we at present unfortunately don't know. Now you may say that it's an act of faith because we will not be able to observe these other big bangs, or these other terms in the wave function. But that's the dilemma that science has been in for a long time. We don't really observe quarks and we will never see the track of a quark. And yet we believe in quarks because the theories that have quarks in them work. And in the same way, if we come to that—and we have not yet come to that—we will believe in these other big bangs or these other terms in the wave function because the theories in which they appear work."[24]

Steve's clear statement reminds me of a poem by Robinson Jeffers, an American poet and brother of a Lick Observatory astronomer. Here are lines from "The Great Wound":

> The mathematicians and physics men
> Have their mythology; they work alongside
>   the truth,

Never touching it; their equations are false
But the things *work.*[25]

Steve Weinberg has more recently added a fascinating sequel to his earlier remarks:[26]

"Any beings like ourselves that are capable of studying the universe must be in a part of the universe in which the constants of nature allow the evolution of life and intelligence. Man may indeed be the measure of all things, though not quite in the sense intended by Protagoras.

"So far, this anthropic speculation seems to provide the only explanation of the observed value of the dark energy. In the standard model and all other known quantum field theories, the dark energy is just a constant of nature. It could have any value. If we didn't know any better we might expect the density of dark energy to be similar to the energy densities typical of elementary particle physics, such as the energy density in the atomic nucleus. But then the universe would have expanded so rapidly that no galaxies or stars or planets could have formed. For life to evolve, the dark energy could not be much larger than the value we observe, and there is no reason for it to be any smaller.

"Such crude anthropic explanations are not what we have hoped for, but they may have to content us. Physical science has historically progressed not only by finding precise explanations of natural phenomena, but also by discovering what sorts of things *can* be precisely explained. These may be fewer than we thought."

Now, as to the efficacy of the multiverse cosmology, to the extent that it is the cosmologists' mythology simply to give a *physical* explanation for the fine-tuning that leads to our existence, I am inclined to write it off as metaphysical fantasy. But should it someday fall into place in our understanding of a grand unified theory that unites quantum mechanics and relativity, then perhaps, as is the case with those unobservable quarks, we would simply have to believe in the plethora of unobservable universes. Nevertheless, I would be unlikely to abandon the idea that those seemingly fine-tuned physical parameters are somehow God-given. My view is of course a final cause, something that can coexist with an efficient cause.

As I was preparing this chapter, I was astonished to discover a resonance with an address given by Fred Hoyle in the University Church in Cambridge in 1959. Hoyle spoke of the stars as serving as gigantic factories in which the whole array of atoms are produced from the simplest atom of all, hydrogen. This lecture was given, I should remind you, very soon after the famous B^{2}FH paper on the astronomical origin of the elements. "If this were a purely scientific question and not one that touched on the religious problem, I do not believe that any scientist who has examined the evidence would fail to draw the inference that the laws of nuclear physics have been deliberately designed with regard to the consequences they produce inside the stars. If this be so, then my apparently random quirks [and here let me interject that what Hoyle calls "random quirks" we would today call fine-tuning]—my apparently random quirks become part of a deep laid scheme. If not, then we are back again to a monstrous sequence of accidents."[27]

I cannot help but think that Hoyle's monstrous sequence of accidents is what the multiverse hypothesis is now being asked to provide.

It is fascinating to note that Hoyle's very next paragraph begins, "There is an interesting similarity between this whole inorganic problem of the origin of the complex atoms and the problem of the origin of life. . . . In both cases we have matter evolving from simpler to more complex forms in accordance with the laws of physics and chemistry. . . . Suppose for a moment you were designing the laws. How much more subtle to make the origin of life implicit in your design—how crude to be obliged to make a gross rectification of your own mistakes!"

Hoyle's remark reminds me of the comment made by the Anglican minister and essayist Charles Kingsley after he read Darwin's *Origin:* "We always knew God was wise, but he was even wiser than we thought, to make a world that could create itself." As already mentioned in my previous chapter, Darwin was so pleased with Kingsley's reaction that he added, anonymously, the cleric's opinion in the second edition of the *Origin.*

Hoyle himself goes on to advance the hypothesis that the laws of science have been designed to promote the origin of life. "Hence if life is part of a deliberate plan so must the origin of the physical conditions be [part of a deliberate plan]. . . . But my object is not to arrive at any complete conclusion. It is to give you a very brief outline of the way that scientific inquiry can be brought into relation with religion." So with this as a pointer, let me turn to a few remarks on the current status of the search for habitable planets beyond our own solar system.

The principle of plenitude asserts that everything that can happen will happen eventually. It was discussed by theologians such as Augustine and Aquinas,[28] and today seems implicit in the idea of multiple universes; it also guides much thinking on the question of intelligent life beyond the solar system. That is to say, if the conditions are right for intelligent life on a distant world, then the principle of plenitude says that eventually there will be intelligent life on that distant world. But, as Paul Davies has reminded us, possibility is not the same as likelihood.

The Copernican system made the Earth into a planet, and by implication the other planets became worlds, possibly inhabited worlds. The Sun became a star, and by implication other stars might have planets, and by the principle of plenitude the planets might be inhabited. Thinking such thoughts about other inhabited planets was one of many reasons why Giordano Bruno got into trouble with the Inquisition. There are scattered allusions in Kepler's writings suggesting that he accepted the notion of inhabitants on other planets. For example, in his *Conversation with Galileo's Sidereus nuncius,* he wrote, "Our moon exists for us on Earth, not the other globes. Those four little moons exist for Jupiter, not for us. Each planet in turn, together with its occupants, is served by its own satellites. From this line of reasoning we deduce with the highest degree of probability that Jupiter is inhabited."[29]

Ninety years later the polymath Christiaan Huygens wrote his *Cosmotheros,* describing a delightful trip through the solar system. He speculated on the state of arts among the inhabitants of Jupiter and Saturn. "If their globe is divided like ours, between Sea and Land, as it's evident it is, we have great reason to allow them the Art of Navigation. . . . If they have Ships, they must have Sails and Anchors, Ropes, Pullies, and Rudders."[30] In other words, there must be hemp on Jupiter for the making of ropes!

Another century later William Herschel, the discoverer of the planet Uranus and an eminent observer and builder of telescopes, wrote in the *Philosophical Transactions* of the Royal Society, "The sun appears to be nothing else than a very eminent, large, and lucid planet.... Its similarity to the other globes of the solar system, with regard to its solidity, its atmosphere, and its diversified surface; the rotation upon its axis, and the fall of heavy bodies leads us on to suppose that it is most probably also inhabited, like the rest of the planets, by beings whose organs are adapted to the peculiar circumstances of that vast globe."[31]

The most extravagant projection came in 1837 from the popular astronomy writer, the Reverend Thomas Dick, whose *Celestial Scenery, or The Wonders of the Planetary System Displayed* includes a table with specific numbers for the populations of planets, asteroids, planetary satellites, and even the rings of Saturn. Basing his calculation on the population density of England [280 persons per square mile, and neglecting the possibility of oceans], he arrived at 6 trillion 967 billion inhabitants on Jupiter, and 8 trillion 141 billion for the rings of Saturn.[32] Plenitude on steroids!

By the end of the century, the Boston Brahmin with his own observatory in Arizona, Percival Lowell, argued for intelligent inhabitants building canals on Mars. Plenitude was then in eclipse. The most widely held view on how planets were formed was the collision theory, and the collision of two stars was seen as an immensely uncommon event in the Milky Way galaxy. Mars remained as a rare possible site for extraterrestrials. But Lowell's Martians were met with widespread skepticism in astronomical circles. Not until the 1950s was the nebular hypothesis revived, with the prospect of millions or more planets within our galaxy. The famous Drake equation provided a framework for estimating not only the number of planets, but the number of technological civilizations in our home galaxy. (The number has proved to be highly variable, depending heavily on the assumed lifetime of a communicating civilization.)

Then, in 1989, my colleague and former thesis student, David Latham, discovered what may prove to be the first planet beyond the solar system. The problem is, it was so unusual that no one could determine whether it was really a planet or just a subdwarf star, and it is taking over two decades to be sure that it really is the first discovered exoplanet. Now Latham is considered the Leif Erikson of exoplanet pioneers. The search for exoplanets has in the last decade burst into a big business among astronomers. I recently consulted with Latham (in May 2014) for the latest number of exoplanet candidates—some thousands, most of which will ultimately be confirmed as planets. For example, of nearly 4,000 candidates announced by the team of astronomers working with NASA's Kepler mission, something like 90 percent will prove to be planets, that is, with properties consistent with theoretical models of the structure and composition of planets. As of May 19, 2014 the number of planets confirmed by the Kepler mission was 966 in systems hosted by some 400 stars.

Part of what is driving the exoplanet search is the hope of finding an earthlike planet with signs

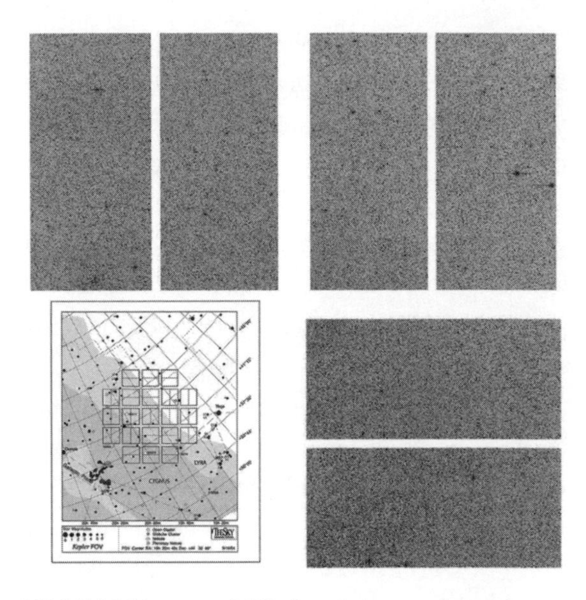

NASA's highly successful Kepler mission was designed to survey a region of the sky near the Milky Way repeatedly, recording the small changes in the brightness of stars when planets intervened, cutting a tiny fraction of the light. Launched in 2009, the mission has monitored the brightness of 145,000 stars. It has found thousands of transit events.

of life. Because oxygen is a very active element that will rust its environment, an oxygen atmosphere will gradually disappear unless it is continually replenished. So far it has been difficult to find a process other than the chemical reactions of life that will sustain an oxygen atmosphere. Hence the detection of an oxygen atmosphere is a critical goal of the exoplanet research program. But researchers would hope to get confirming evidence, such as a methane signature, as well.

Once again the question of belief enters the game. A fundamental belief of many of my colleagues is the principle of mediocrity. Despite the fact that the human brain is the most complex thing known to science, it seems somehow too human-centric to think we are tops in the universe. In fact, the human brain is sufficiently complex to speculate that it is *not* the most complex object in the universe. These various evaluations of our place in the universe, from the principle of plenitude to the principle of mediocrity, belong to a system of philosophical belief, a magisterium, that is not part of science, but obviously it overlaps with the magisterium that encompasses the nature of science.

I would personally not be shocked if the spectroscopists find an oxygen signal. For a billion years imaginary scientists on a distant world observing our solar system from afar could have found a weak oxygen signal from our planet, but only bacteria would have caused it. It would be fabulous for us to find such a signal from afar, to know that our universe is somehow designed to be favorable to the origin of life. Habitable, yes, but not necessarily inhabited, unless bacteria count as inhabitants. The oxygen signal alone would not, however, give us a clue about the existence of intelligent life. But it will surely help mold the magisterium of belief.

As I now conclude this third excursion into the nature of science, I suppose I should specifically answer the question of whether Fred Hoyle was right. The answer is, of course, sometimes! And one of the ideas that sometimes came up in his reflections was the notion of a superintelligence designing it all. As I said earlier, I accept as a final cause that the physical constants have been fine-tuned to make intelligent life in the universe possible and that this is evidence for the planning and intentions of a Creator God. But this evidence does not stand alone. The mere fact that the universe is comprehensible to our minds is also powerful evidence for a superintelligent designer. In this

Fred Hoyle was close to being a religious atheist. Hoyle was very far from the traditional Christianity so brilliantly defended by his contemporary and fellow BBC commentator, C. S. Lewis, with his book *Mere Christianity.* And yet by the deep way he understood the nature of science, Hoyle was not that far from final causes.

And finally, I hope that in these chapters I have persuaded you that what is accepted today as science is commonly colored by personal beliefs, including our religious or our antireligious sentiments. If someone tells you that evolution is atheistic, be on guard. If someone claims that science tells us we are here by pure chance, take care. And if someone declares that magisteria do not overlap, just smile smugly and don't believe it.

Notes

1. Was Copernicus Right?

1. Owen Gingerich, *The Book Nobody Read: Chasing the Revolutions of Nicolaus Copernicus* (New York: Walker, 2004), 135.
2. Arthur Koestler, "In Memory of Franz Hammer," in *Vistas in Astronomy*, vol. 18, eds. Arthur Beer and Peter Beer (Oxford: Pergamon Press, 1975), 947–949.
3. Arthur Koestler, *The Sleepwalkers: A History of Man's Changing Vision of the Universe* (New York: Macmillan, 1959), 191.
4. Gingerich, *The Book Nobody Read,* note 1, 255.
5. G. J. Rheticus, *Narratio prima,* in *Three Copernican Treatises,* Edward Rosen, trans. (New York: Octagon Books, 1971), 165.
6. Owen Gingerich, "Erasmus Reinhold and the Dissemination of Copernican Theory," in Gingerich, *The Eye of Heaven: Ptolemy, Copernicus, Kepler* (New York: American Institute of Physics, 1993), 221–251.
7. Christopher Graney, "Science Rather Than God: Riccioli's Review of the Case for and against the Copernican Hypothesis," *Journal for the History of Astronomy* 43 (2012): 215–225.
8. Adrian Wilson, *The Making of the Nuremberg Chronicle* (Amsterdam: Nico Israel, 1976), 76, 195–205.

9. Galileo Galilei, *Dialogue Concerning the Two Chief World Systems,* trans. Stillman Drake (Berkeley: University of California Press, 1953), 328.
10. Stillman Drake, *Discoveries and Opinions of Galileo* (Garden City, NY: Doubleday, 1957), 165.
11. Richard Kluger, *Simple Justice: The History of Brown v. Board of Education and Black America's Struggle for Equality* (New York: Vintage Books, 1977).
12. Owen Gingerich, "The Galileo Sunspot Controversy: Proof and Persuasion," *Journal for the History of Astronomy* 34 (2003): 77–78.
13. Owen Gingerich, "The Galileo Affair," *Scientific American* 246 (August 1982), 133–143.
14. Robert Hooke, *An Attempt to Prove the Motion of the Earth from Observations* (London: John Martyn, 1674), 1, 3.
15. Ibid., 25.

2. *Was Darwin Right?*

1. Gallup poll, February 11, 2009 (Darwin's birthday).
2. Charles Darwin, *Journal of Researches into the Geology and Natural History of the Various Countries Visited by H.M.S. Beagle,* 2nd ed. (London: Ward Lock & Co., 1845), 173.
3. Richard Darwin Keynes, ed., *The Beagle Record* (Cambridge: Cambridge University Press, 1979), 185 (quoting from Darwin's diary).
4. Ibid., 128, Darwin to Henslow, April 11, 1833; *The Correspondence of Charles Darwin,* vol. 1 (Cambridge: Cambridge University Press, 1985), 306.
5. Ibid., 263 (quoting from Darwin's diary).

6. Ibid., 263–264, Darwin to his sister Caroline, March 10, 1835; *Correspondence,* vol. 1, note 4, 434.
7. Adrian Desmond and James Moore, *Darwin* (London: Michael Joseph, 1991), 694, *Journal* 1839, 474–475.
8. Darwin, *Journal and Remarks, 1832–1836* (London: Henry Colburn, 1839), 474.
9. Darwin to his sister Susan, August 4, 1836, *Correspondence,* Vol. 1, note 4, 503.
10. Nora Barlow, ed., *The Autobiography of Charles Darwin* (London: Collins, 1958), 120.
11. Darwin to Gray, July 20, 1857, *Correspondence,* vol. 6, 432.
12. Sara Joan Miles, "Charles Darwin and Asa Gray Discuss Teleology and Design," *Perspectives on Science and Christian Faith,* 53 (2001): 196–201.
13. Darwin to Gray, September 5, 1857, *Correspondence,* Vol. 6, note 11, 445–449.
14. James E. Strick, *Sparks of Life: Darwinism and the Victorian Debates over Spontaneous Generation* (Cambridge: Harvard University Press, 2000).
15. Neil Shubin, *Your Inner Fish* (New York: Pantheon Books, 2008), 23–25.
16. Ian Tattersall and Jeffrey Schwartz, *Extinct Humans* (New York: Westview Press, 2000), 244.
17. G. David Poznik, et al., "Sequencing Y Chromosomes Resolves Discrepancy in Time to Common Ancestors of Males versus Females," *Science* 341 (2013): 562–565.
18. Francisco J. Ayala, "The Myth of Eve: Molecular Biology and Human Origins," *Science* 270 (1995): 1930–1936.
19. Alan R. Rogers and Henry Harpending, "Population Growth Makes Waves in the Distribution of Pairwise

Genetic Differences," *Molecular Biology and Evolution* 9 (1992): 552–569.

20. Wentzel van Huyssteen, *Alone in the World: Human Uniqueness in Science and Theology* (Grand Rapids: Eerdmans, 2006), 234.
21. John Paul II, *Message Delivered to the Pontifical Academy of Sciences, October 22, 1996.*
22. Asa Gray, *Darwiniana: Essays and Reviews Pertaining to Darwinism* (New York: Appleton, 1876), 158; in the reprint (Cambridge: Harvard University Press, 1963), 130.
23. Charles Kingsley, *Westminster Sermon* (London: Macmillan, 1874), xxvii.
24. Kingsley to Darwin, November 18, 1859, *Correspondence,* vol. 7, 379–380.
25. Darwin to Gray, November 26, 1860, *Correspondence,* vol. 8, 496.
26. Darwin to Gray, September 17, 1861, *Correspondence,* vol. 9, 267–268.
27. Mortimer J. Adler, "Natural Theology, Chance, and God," *The Great Ideas Today* 1992 (Encyclopedia Britannica, Chicago, 1992), 287–301; Owen Gingerich, "Response to Mortimer J. Adler," 302–304.
28. George Gaylord Simpson, *The Meaning of Evolution,* rev. ed. (New Haven: Yale University Press, 1967), 345.
29. Jon D. Miller, Eugenie C. Scott, Shinji Okamoto, "Public Acceptance of Evolution," *Science* 313 (2006): 765–766.
30. Freeman Dyson, *Disturbing the Universe* (New York: Harper and Row, 1979), 250.

3. Was Hoyle Right?

1. Owen Gingerich, "In the Orbit of Copernicus," *American Scholar* 80, no. 3 (Summer 2011): 43–49. This essay was originally titled "Miracle on the Warsaw Express, or How I Got Invited to Copernicus' Funeral."
2. Simon Mitton, *Conflict in the Cosmos: Fred Hoyle's Life in Science* (Washington, DC: Joseph Henry Press, 2005), with a foreword by Owen Gingerich.
3. "Memoranda by David Gregory," *The Correspondence of Isaac Newton*, vol. 3 (Cambridge: Cambridge University Press, 1961), 334–336.
4. Michael Hoskin, "Newton, Providence and the Universe of Stars," *Journal for the History of Astronomy* 8 (1977): 77–101.
5. Mario Livio, *Brilliant Blunders: from Darwin to Einstein*, (New York: Simon and Schuster, 2013).
6. Fred Hoyle, *The Nature of the Universe*, (New York: Harper, 1950), 119.
7. Ibid., 139, 141.
8. Dorothy Sayers, *The Listener*, November 9, 1950, 496–97.
9. Helge Kragh, *Cosmology and Controversy*, (Princeton: Princeton University Press, 1960), 257.
10. George Gamow, "The Role of Turbulence in the Evolution of the Universe," *Physical Review* 86 (1952): 251.
11. Rodney Holder, *Big Bang, Big God* (London: Lion Hudson, 2013), 12.
12. Simon Mitton, *Conflict in the Cosmos*, 223–224. This section relies on Ward Whaling's account in an oral history interview posted online by the Caltech Archives.

13. Geoffrey Burbidge, "Hoyle's Role in B^2FH," *Science* 319 (2008): 1484.
14. Fred Hoyle, "The Universe: Past and Present Reflections," *Engineering and Science* (November 1981): 8–12, especially 12.
15. Fred Hoyle, *The Origin of the Universe and the Origin of Religion* (Wakefield, RI: Moyer Bell, 1993), 83.
16. Francis Bacon, *On the Advancement of Learning* (Chicago: Great Books of the Western World, 1952), vol. 30, 43.
17. Ibid., 111.
18. Martin Gardner, *Are Universes Thicker Than Blackberries?* (New York: Norton, 2003).
19. George F. R. Ellis, "Opposing the Multiverse," *Astronomy and Geophysics* 49, no. 2 (April 2008): 33.
20. Bernard Carr, "Defending the Multiverse," *Astronomy and Geophysics* 49, no. 2 (April 2008): 37.
21. This paragraph and the following one are quoted almost verbatim from Professor Davis's remarks following my lecture at Gordon College.
22. Lord Salisbury, "Unsolved Problems of Science," *Popular Science Monthly* 46 (November 1894): 94.
23. Quoted by Joseph Larmor in the article "Aether" in the eleventh edition, volume 1, of the *Encyclopaedia Britannica* (London and New York, 1910), 292. Encyclopaedia Britannica.
24. Steven Weinberg, "An Exchange between Steven Weinberg and John Polkinghorne," *Annals of The New York Academy of Sciences*, 950 (2001), 190.
25. Robinson Jeffers, "The Great Wound," in *The Beginning and the End* (New York: Random House, 1963).

26. Steven Weinberg, "Physics: What We Do and Don't Know," *New York Review of Books* (November 7, 2013): p. 88.
27. Fred Hoyle, "Address Delivered in the University Church, Cambridge," in Merveyn Stockwood, ed., *Religion and the Scientists* (London: SCM Press, 1959), 64.
28. Arthur Lovejoy, "The Principle of Plenitude and the New Cosmography," in Lovejoy, *The Great Chain of Being*, ch. 4 (Cambridge, MA: Harvard University Press, 1936), 99–143.
29. Johannes Kepler, *Dissertatio*, quoted in Steven J. Dick, *Plurality of Worlds* (Cambridge: Cambridge University Press, 1982), 77.
30. Christiaan Huygens, *The Celestial Worlds Discover'd; or, Conjectures Concerning the Inhabitants, Plants and Productions of the Worlds in the Planets* (London, 1698), 82–83.
31. William Herschel, "On the Nature and Construction of the Sun and Fixed Stars," *Philosophical Transactions* (1795), 46–72, reprinted in *The Scientific Papers of Sir William Herschel*, vol. 1 (London, 1912), 479.
32. Cited in Michael J. Crowe, *The Extraterrestrial Life Debate, 1750–1900* (Cambridge: Cambridge University Press, 1986), 199.

Acknowledgments

The Herrmann Lectures that form the basis of *God's Planet,* given at Gordon College in Wenham, Massachusetts in October of 2013, were established in honor of Robert Herrmann, the long-time executive officer of the American Scientific Affiliation. I have known and admired Bob Herrmann for several decades, and we have traveled thousands of miles together visiting ASA chapters throughout America. This has been my opportunity to thank Bob for his friendship and the interesting discussions we have had along the way.

I gratefully acknowledge the Center for Faith and Inquiry at Gordon College for organizing and hosting the event, particularly Thomas Albert Howard and M. Ryan Groff, and also the Herrmann Lecture committee. They had received a grant from the John Templeton Foundation to make the lectures possible. I am pleased to acknowledge that each lecture was reinforced by a response given each afternoon, by Randy Isaac, Steve Alter, and Edward Davis, and I specially thank Ted Davis for allowing me to incorporate part of his thoughtful reflections into the final text of my third lecture.

Friends and colleagues have read various portions of my text, with very helpful insights, enhancements, and corrections. Among them are Michael Hoskin, David Latham, Simon Mitton, Howard Smith, Rodney Holder, Dennis Venema, and Jeff Hardin. Janet Browne, Martin Rees, and Philip Gingerich have been of indispensable help with respect to key illustrations, and Nathan Sanders has been extraordinarily efficient in drafting the final versions of the diagrams.

I have been tremendously encouraged in assembling this manuscript by the enthusiasm and encouragement from Michael Fisher of the Harvard University Press. And finally, I am grateful beyond words for the support and patience of my wife Miriam in this whole process.

Index